Boolean Algebra
for
Computer Logic

by
Harold E. Ennes

Howard W. Sams & Co., Inc.
4300 WEST 62ND ST. INDIANAPOLIS, INDIANA 46268 USA

International Standard Book Number: 0-672-21554-3
Library of Congress Catalog Card Number: 78-62351

Printed in the United States of America.

Preface

The purpose of this book is to provide a basic introduction to Boolean algebra as applied to the design of computer circuits. Since Boolean algebra is a modification of conventional algebra, a general understanding of elementary algebra is a prerequisite. This text contains only a very brief review of factoring and the associative, distributive, and commutative rules. No other prerequisite is required, but an understanding of elementary electronics is helpful though not a must.

Since the Boolean technique involves variables that have only two possible "levels" (0 and 1), the binary number system is covered first. This is followed by symbolic logic and truth tables as derived from typical logic statements. The text then takes up the study of algebraic techniques for optimal design of logic circuits. Venn diagrams and Karnaugh maps are also covered as means of circuit simplification. The final chapter consists of solved problems over the material in the preceding chapters.

The format of this book is primarily self-instructional, with exercises at the end of each chapter and solutions to the exercises in the Appendix of the book. The text is, however, also suitable for use in a formal course in fundamentals of logic design for students who have had no previous exposure to Boolean algebra.

HAROLD E. ENNES

Contents

CHAPTER 3

CHAPTER 4

CHAPTER 5

CHAPTER 6

APPENDIX

CHAPTER 1

Introduction to the Boolean-Binary Relationship

In 1847, a mathematician named George Boole (1815–64) demonstrated an exciting fact: that when logical statements are given simplified symbolic forms, these symbols can be manipulated very much like algebraic symbols. This type of symbolic logic is called *Boolean algebra,* and the terms "symbolic logic" and "Boolean algebra" mean one and the same thing. In 1854, George Boole collected and reviewed all of his work in his most famous treatise, *The Laws of Thought.* All modern digital computer logic operations are based upon the laws proposed in that work.

1-1. THE BASIC BOOLEAN PREMISE

First of all, Boolean arithmetic recognizes only two values or levels, which may be represented by 1 and 0. In this case, one and zero do not necessarily represent numbers; the "1" can mean "true" and the "0" can mean "false." Or the one and the zero can be arbitrarily assigned to anything—Jack and Jill, white and black, etc. In any case, the one is always assumed to be a higher level than the zero in any electronic circuitry.* Thus a closed switch can represent maximum current in a load (logic 1) and an open switch can represent no current (logic 0). And, to keep your perspective, ground (0 volts) can be "high" if logic 0 is a negative voltage relative to ground.

In conventional arithmetic, there are four basic operations: add, subtract, multiply, and divide. There are only three basic operations in Boolean algebra: AND, OR, and NOT (inversion). Although these do not sound mathematical, you will find a direct correlation between logical statements and these three basic operations used to describe them.

Constants

Recall that a constant is a value, quantity, etc., that has a fixed meaning. For example, the number 4 always means the same thing. Thus in conventional algebra many possible constants exist, such as all integers and all fractions.

Boolean algebra, however, has only two possible constants: 0 and 1. These constants describe the two allowable voltage levels on a wire, or current levels carried by the wire. In electronic parlance, a "pulse" is a logical 1, and no pulse is a logical 0. The magnitude of voltage or current is arbitrary; logical 1 simply says the voltage or current is positive relative to that of logical 0.

Variables

A variable is a value, quantity, etc., which can *change* to the value of any constant in the system. Thus at some given time, the variable is at the value of a constant, but at an earlier time or later time it may take on the value of any other constant. Since there are only two constants in the Boolean system, the variable can assume only one of the two values.

The variable, whose unspecified value is either a 0 or a 1, is denoted by letters A, B, C, . . . , X, Y, Z, either capitalized or

---

* For "positive" logic, which is assumed in this book.

small. A variable may be inverted, in which case a bar is placed over it:

If $A = 1, \bar{A} = 0$.
If $A = 0, \bar{A} = 1$.

This is simply another way of saying that since there are only two possible levels, if the level is NOT a 1, it must be a 0. Or, if the level is NOT 0, it must be 1. Thus:

$$\bar{1} = 0 \text{ and } \bar{0} = 1.$$

Just remember at this time that the noninverted form of the variable has no bar over it. The NOT, or inverted form, of the variable has a bar over it.

Operations

Although Boolean algebra has many things in common with conventional algebra, it does have differences. Although the two possible values of 0 and 1 do not necessarily represent numbers, let's temporarily consider them as such. If you take the product $1 \cdot 1 \cdot 1 \cdot 1$ (one times one times one times one) you have a product of 1. This is said to equal 1, and the $=$ sign has conventional meaning of equality.

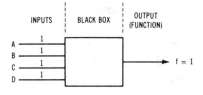

Fig. 1-1. If A and B and C and D are all 1s, the function f (output) is a 1.

A computer logic element is a "black box" which has two or more inputs, with a single output that is a direct function of the inputs. Fig. 1-1 shows such a box with four inputs, all of which are "highs" or logical ones. If the function (output) is a 1, all we know at the present time is that when all inputs are high, the function (f) is also high (an output occurs).

Now consider the product $1 \cdot 0 \cdot 1 \cdot 1$. You will recall that 1 times

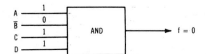

Fig. 1-2. If any one or all inputs
are at logic 0, then f = 0
(no output occurs).

0 is always 0, and 0 times 1 is always 0. Thus the product of
$1 \cdot 0 \cdot 1 \cdot 1 = 0$ (no output occurs).

Fig. 1-2 shows the black box of Fig. 1-1 with a 0 on the B
input ($\bar{B}$) and reveals that the output (f) is now 0. Further
tests reveal that if *any one* of the inputs is a 0, the function
(output) is also 0. In symbolic logic, this is termed an AND circuit. All inputs must be simultaneously high for an output to
occur. If any one input is 0, or if all inputs are 0, the output will
be 0. Note that this says A AND B AND C AND D (written ABCD)
must all be high (no bars) for a "1" output to occur. In Fig. 1-2,
$A\bar{B}CD$ (read A AND B NOT AND C AND D) = 0. Note, therefore,
that the multiplication sign in Boolean algebra is Boolean operation AND. Usually, as in conventional algebra, the multiplication
(times) sign is simply deleted.

Now let's consider the sum (add) operation. Suppose you have
1+1+1+1. In conventional arithmetic, this would equal the
quantity 4. The conventional number system has a base of ten,
and consists of numbers 0–9. When you reach the sum of 9, you
change that column to 0 and start a second column (to the left)
with a 1, to obtain the quantity 10 (ten).

A number system that consists of only 0 and 1 is termed a
binary (base 2) system. Since 1 is the highest number in the
system, 1+1 = 0 and a 1 is carried to the left as: 10. This does
not read "ten," but it is read as "one-zero." You will get ac-
quainted with the binary system in this chapter.

Assume for the moment you have the black box of Fig. 1-3.
You find that if either or both inputs are high, an output occurs.
This says that if A or B (or A or B or C or D, etc.) is high, an
output occurs. Since an output occurs if either one or all inputs

Fig. 1-3. If a 1 exists at any input,
the function f (output) is a 1.

are high, it is termed an *inclusive* OR Boolean operation. In Boolean notation, the connective OR is signified by the sum $(+)$ sign. Thus $A + B + C + D$ reads A OR B OR C OR D. This says that an output will occur $(f = 1)$ if any one or more inputs is 1. If all inputs are 0, no output occurs $(f = 0)$.

How a combination of AND and OR black boxes is used to obtain binary sums and other mathematical as well as logic statement operations will be revealed after you learn the technique of the binary system. To summarize thus far:

AB means A AND B.

$A + B$ means A OR B.

$\overline{AB}$ means NOT A AND B.

A bar over a letter or group of letters always indicates the inverted form of the indicated letter or quantity.

1-2. THE POWER OF A NUMBER (EXPONENT)

In our conventional decimal system, $10^1 = 10$ and $10^{-1} = 0.1$. Now $10^0 = 10^{1-1} = 10^1 \times 10^{-1} = 10 \times 0.1 = 1$. Thus $10^0 = 1$. *Any* number raised to the zeroth power is equal to 1. So $2^1 = 2$ and $2^{-1} = 0.5$ and $2^0 = 1$.

Table 1-1 reviews the decimal (base 10) system for four digits.

Table 1-1. Decimal 1 System for Four Digits

	Positional Weight			
	10^3	10^2	10^1	10^0
Equivalent	1000	100	10	1
	3	0	0	0
		2	0	0
$3248 =$			4	0
				8
	$= 3$	2	4	8
	↑ MSD			↑ LSD
Showing That $3248 = 3000+200+40+8$				

The 10^0 column contains units of ones, the 10^1 column is units of tens, 10^2 is units of hundreds, and 10^3 is units of thousands. Thus the number $3248 = 3000 + 200 + 40 + 8$. The power is normally termed the *positional weight* of the number. Thus a 6 in the 10^3 column would represent 6000 while a 6 in the 10^0 column represents only 6. The least significant digit (lsd) is in the 10^0 column while the most significant digit (msd) is in the 10^3 column in this table.

Table 1-2 reveals the corresponding binary system for four bits. ("Bits" is a contraction of *bi*nary digi*ts*). Here, instead of ascending powers of ten, we have ascending powers of two. Only zeroes and ones are used. A 1 in the 2^0 column (least significant bit) is equal to 1. A 1 in the 2^1 column is equal to 2. A 1 in the 2^2 is equal to 4. A 1 in the 2^3 column is equal to 8.

Table 1-2. Binary Table for Four Bits

Binary Weight	2^3	2^2	2^1	2^0	
Decimal Equivalent	8	4	2	1	Decimal Count
	0	0	0	0	0
	0	0	0	1	1
	0	0	1	0	2
	0	0	1	1	3
	0	1	0	0	4
	0	1	0	1	5
	0	1	1	0	6
	0	1	1	1	7
	1	0	0	0	8
	1	0	0	1	9
	1	0	1	0	10
	1	0	1	1	11
	1	1	0	0	12
	1	1	0	1	13
	1	1	1	0	14
	1	1	1	1	15
	MSB			LSB	

To convert any 4-bit binary number to its equivalent decimal value, count from right to left and give each 1 its equivalent decimal value from the "weight" of its given position. For example, binary 0110 is simply $2 + 4 = 6$. Binary 1001 is simply $1 + 8 = 9$. Binary $1111 = 1 + 2 + 4 + 8 = 15$.

Going Beyond Four Bits

Table 1-3 lists the positive powers of 2 for values up to 4096 decimal equivalent. Note that each additional power (weight) doubles the decimal equivalent value. Thus Tables 1-1 and 1-2 can be expanded to as many powers as required for a particular application. For example, binary $100101 = 32 + 4 + 1 = 37$. Binary $1100101 = 64 + 32 + 4 + 1 = 101$. Simply count the spaces to the left and give the digit the appropriate decimal equivalent, then add. In counting from right to left, just say 1-2-4-8-16-32-64-128-256, etc.

The Modulus

The odometer (mileage indicator) in your automobile normally has a decimal readout capacity of 99,999.9 miles. When you travel another 1/10 mile, the odometer "resets" or "clears" to 00,000.0, which actually represents 100,000 miles. Therefore,

Table 1-3. Positive Powers of 2

Power of 2	Decimal Equivalent
2^0	1
2^1	2
2^2	4
2^3	8
2^4	16
2^5	32
2^6	16
2^7	128
2^8	256
2^9	512
2^{10}	1024
2^{11}	2048
2^{12}	4096

for this device, the *modulus* is 100,000. The modulus of any counter is the fixed capacity of the counter *plus one* increment of the least significant digit (lsd).

Note from Table 1-2 that the 4-bit binary counts from 0 to 15. This is 16 "levels," including zero, and 16 is the modulus ($2^4 = 16$). Another way to look at this is to realize that the maximum decimal count is the modulus minus 1.

1-3. ADDING BINARY ZEROES AND ONES

Start by writing down four zeroes:

$$0000 \text{ (make this equal to 0)}$$
$$+\underline{\quad 1} \text{ (add 1)}$$
$$0001 \text{ (this equals 1 in the decimal system)}$$

Now remember that decimal 9 is the highest number used in base 10 arithmetic. Thus when you reach 9, the addition of 1 changes 9 to a 0, and the 1 is carried to the "tens" column.

In binary, the highest number used is 1. So consider the addition of another 1 to 0001:

$$0001$$
$$+\underline{\quad 1}$$

When 1 is directly under 1 in the binary system, the sum is 0 and a 1 is carried to the next column to the left. If there is already a 1 in that column, the sum is again 0 and the 1 carries to the left again, etc. So:

$$0001$$
$$+\underline{\quad 1}$$
$$0010$$

This equals 2 in the decimal system because you have accumulated two pulses (two ones) and the "1" is in the second space to the left under the column 2 (Table 1-2). Now add another 1:

$$0010$$
$$+\underline{0001}$$
$$0011$$

(This is $1 + 2 = 3$ in the decimal system.) Add another 1:

$$0011$$
$$+ \quad 1$$
$$\overline{0100}$$

This is 4 in the decimal system, since the "1" is now in the third space to the left ($2^2 = 4$) as in Table 1-2.

When you continue this process to 1111, you have developed the 4-bit binary table of Table 1-2.

For practice, add 0111 with 1110 as in Fig. 1-4. Row 1 contains the augend and row 2 the addend. Row 3 shows the addition of 0 and 1, which equals 1 with no carry. Row 4 adds 1 and 1, giving 0 and carry 1. In row 5, column C, you already have two 1s with a previous carry of 1. In this case, you must put down your previous carry as shown; then you still carry 1 to the next column. In row 6, column D, you have $1 + 1 = 0$ and carry 1. In row 7, you simply record the carried 1.

ROW	COLUMN D C B A	
1	0 1 1 1	AUGEND
2	+ 1 1 1 0	ADDEND
3	1	NO CARRY
4	0	CARRY 1
5	1	1 CARRIED AND CARRY 1
6	0	CARRY 1
7	1	1 CARRIED
8	1 0 1 0 1	BINARY TOTAL

Fig. 1-4. Addition of binary numbers 0111 and 1110.

Check this in decimal form. You know that binary 0111 (read zero-one-one-one) is equal to decimal 7 ($1 + 2 + 4 = 7$). Also that $1110 = 14$. So $7 + 14 = 21$, and binary $10101 = 21$.

Table 1-4 reviews the binary table for addition. Memorize these rules.

Fractions

You have already noted that the digits to the left of the binary point are coefficients of increasing *positive* powers of 2, with 2^0 adjacent to the binary point. The digits on the right of the point are coefficients of increasing *negative* powers of 2, with 2^{-1} adjacent to the point.

Table 1-4. Rules for Binary Addition

Binary A + B + Prev Carry	= Sum	New Carry
0 + 0 + 0	= 0	0
0 + 0 + 1	= 1	0
0 + 1 + 0	= 1	0
0 + 1 + 1	= 0	1
1 + 0 + 0	= 1	0
1 + 0 + 1	= 0	1
1 + 1 + 1	= 1	1

See Table 1-5 and note that a 1 immediately to the right of the decimal point would be equal to $\frac{1}{2}$ or 0.5, while a 1 in the fourth

Table 1-5. Binary Numbers With Fractions

Binary Weight	2^3	2^2	2^1	2^0	Point	2^{-1}	2^{-2}	2^{-3}	2^{-4}
Decimal Equivalent	8	4	2	1	.	1/2 (0.5)	1/4 (0.25)	1/8 (0.125)	1/16 (0.0625)

place to the right would be equal to $\frac{1}{16}$, or 0.0625. Thus 0000.1 in binary is $\frac{1}{2}$ or 0.5 in decimal, while 0000.0001 = $\frac{1}{16}$ or 0.0625 in decimal form. For example, the decimal equivalent of binary 0101.01 is:

Binary		Decimal
0×2^3	=	0
1×2^2	=	4
0×2	=	0
1×2^0	=	1
0×2^{-1}	=	0
1×2^{-2}	=	0.25
0101.01	=	5.25

In practice, of course, you simply disregard all zeroes and note the binary weight of the 1 for each position. When you get familiar with binaries, you would look at 0101.01 and say: 4+1+0.25 =

5.25. In a 4-bit binary you simply have (from left to right) decimal values 8-4-2-1 plus any fractional quantity.

1-4. SUBTRACTING BINARY ZEROES AND ONES

Binary subtraction on paper is no different than decimal subtraction; the same "borrow" rules apply. For example:

$$
\begin{array}{r}
1111 \quad (\text{decimal } 15) \\
-0110 \quad (\text{decimal } 6) \\
\hline
1001 \quad (\text{decimal } 9)
\end{array}
$$

In this case binary subtraction is simpler than the decimal form since 15−6 requires "borrowing" the 1.

Now suppose you have:

$$
\begin{array}{r}
44.00 \\
-10.25 \\
\hline
33.75
\end{array}
$$

Observe here that you borrowed twice until a number existed (third digit from right) to borrow from. The subtraction 44−10.25 in binary form is:

$$
\begin{array}{r}
101100.00 \quad (\text{decimal } 44 \;) \\
-001010.01 \quad (\text{decimal } 10.25) \\
\hline
100001.11 \quad (\text{decimal } 33.75)
\end{array}
$$

Making the first digit the one to the extreme right, the procedure is as follows:

1st digit: 0−1 is a difference of 1 and borrow 1.

2nd digit: No 1 exists to borrow yet. So 0 in top row becomes 1 and 1−0 = 1 and borrow 1 carried over.

3rd digit: Still no 1 to borrow from. So 0 in top row becomes 1 and 1−0 = 1 and borrow 1 carried over.

4th digit: Still no 1 to borrow so 0 in top row becomes 1 and 1−1 = 0 with borrow 1 carried over.

5th digit: Now a 1 exists to borrow, so it becomes a 0 and 0−0 = 0. No carry.

The remainder is self-explanatory.

Table 1-6. Rules for Binary Subtraction

A — B — Previous Borrow	= Difference	New Borrow
0 — 0 — 0	0	0
0 — 0 — 1	1	1
0 — 1 — 0	1	1
0 — 1 — 1	0	1
1 — 0 — 0	1	0
1 — 0 — 1	0	0
1 — 1 — 0	0	0
1 — 1 — 1	1	1

Here is another example:

$$\begin{array}{r} 0110 \\ -0011 \\ \hline 0011 \end{array} \quad \begin{array}{l} \text{(decimal 6)} \\ \text{(decimal 3)} \\ \text{(decimal 3)} \end{array}$$

1st digit: 0−1 = difference of 1 and borrow 1.

2nd digit: 1 borrowed from top row (becomes 0) and 0−1 is a difference of 1 and borrow 1.

3rd digit: 1 borrowed from top row (becomes 0) and 0−0 = 0.

4th digit: 0−0 = 0.

Table 1-6 tabulates the rules for binary subtraction.

When we progress to the "fascinating complement" (next section) you will find binary subtraction much simplified.

1-5. HANDLING POSITIVE AND NEGATIVE BINARIES

You have positive and negative values in binary notation just as in base 10 arithmetic. It is time now to consider the binary scheme of manipulating either polarity in calculations. You will find binary subtraction by "complements" much simpler than the straight arithmetical method described above.

The Fascinating Complement

A *complement* is that quantity or amount which, when added to a given quantity, completes a whole. For example, in Fig. 1-5,

18

if you want to complete an angle of 90°, the arc BD is the *complement* of arc AB, and the angle BCD is the complement of the angle ACB. Another way of saying this is that, for an angle of 90° as a whole, the complement of 30° is 90°−30° = 60°.

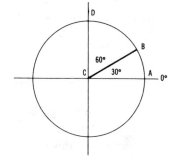

Fig. 1-5. Complement of an angle.

For decimal numbers the complement of 460 is 1000−460 = 540. The complement of 25 is 100−25 = 75. Note in each case that the arithmetical complement is actually the *difference* between a number and the power of the base next in series. When you complement a positive number, it becomes a *negative* value, and the complement is a *positive* number. Synonyms for the "complement" are: inversion, reversal, the opposite, the reverse, the inverse, the converse.

In conventional arithmetic, suppose you subtract 235 from 485:

$$
\begin{array}{rl}
485 & \text{(minuend)} \\
-235 & \text{(subtrahend)} \\
\hline
250 & \text{(answer)}
\end{array}
$$

You can do the same thing by adding rather than subtracting if you complement the subtrahend and continue as in normal addition:

$$
\begin{array}{rl}
485 & \text{(minuend)} \\
+765 & \text{(complement of subtrahend)} \\
\hline
1250 & \text{(answer with 1 overflow)}
\end{array}
$$

In complementary arithmetic, the most significant digit is not a part of the answer numerically (1 in the above example) and the answer (250) remains.

Let's see what happens if you have a result which is a negative number rather than a positive number. Suppose you have 235–485. Normally you simply subtract the smaller number from the larger number and affix the sign of the larger number to the answer. Hence −485 + 235 = −250.

In complementary arithmetic:

$$
\begin{array}{ll}
235 & \text{(minuend)} \\
+515 & \text{(complement of a negative number is positive)} \\
\hline
750 & \text{(complement of 250)}
\end{array}
$$

Hence recomplementing 750 gives −250 (answer). The complement of a complement is the original number.

Simply note from this example that when a resultant is a negative number, it will be indicated by the appearance of the complement form in the answer. The answer is then recomplemented to get the correct negative value.

Binary Complements

Binary complements are much simpler than conventional complementary arithmetic complements. To complement a binary number, simply invert the number (change all zeroes to ones and all ones to zeroes), as you would do in a single electronic polarity-inverter stage. This is called the *ones complement*. Then add 1 for the modulus of the counter (this is called the *twos complement*). For example, to complement binary 0010 (decimal 2):

$$
\begin{array}{ll}
1101 & \text{(inverted 0010: ones complement)} \\
+\quad 1 & \text{(add 1)} \\
\hline
1110 & \text{(twos complement of 0010)}
\end{array}
$$

Remember that the modulus for a 4-bit number is $15+1 = 16$ ($2^4 = 16$). Thus the complement of 2 in this system is $16−2 = 14$, which is binary 1110 as derived above.

Now consider the addition of 0110 and −0100 (decimal 6−4). Simply complement the negative number (gives a positive number), then add the result to the positive number as:

```
    1011   (inverted 0100: ones complement)
+      1   (add 1)
    1100   (complement of 0100: twos complement)
```

Then add this complement to 0110:

```
     0110
    +1100
    10010   (answer)
```

Now you discard the overflow digit (msd) and replace it with either a plus or minus sign. If the msd is 1, the sign is plus. If the msd is zero, the sign is negative. In the above example, the msd is 1, so the answer is +0010, or decimal 2. This satisfies the condition of 6−4 = 2.

When a negative number is added to a smaller positive number, the overflow (msd) is always 0, and the answer is negative. When the result is negative, the number must be *recomplemented* to obtain the correct solution. For example, add −1001 (decimal −9) to 0010 (decimal 2). First find the complement of 1001:

```
    0110   (ones complement)
       1
    0111   (twos complement)
```

Now add this to 0010:

```
      0111
    + 0010
     01001
```

Since the msd is a 0, we discard it and write a negative sign to get −1001. Now to *recomplement:*

```
    0110   (inverted)
+      1   (add 1)
    0111   (negative answer)
```

Thus the answer is −0111. This is decimal −7, which results from adding −9 to +2.

Binary subtraction by complements is exactly the same as

addition of positive and negative binary numbers. For example, subtract 0011 (decimal 3) from 1111 (decimal 15). First find the complement of 0011:

$$
\begin{array}{r}
1100 \quad \text{(inverted: ones complement)} \\
+ \quad 1 \\
\hline
1101 \quad \text{(twos complement)}
\end{array}
$$

Then add:

$$
\begin{array}{r}
1111 \\
+ \quad 1101 \\
\hline
11100 \end{array} = +1100 = \text{decimal 12}
$$

1-6. CONVERTING FROM DECIMAL TO BINARY

You have discovered from previous study how easy it is to convert binary numbers to decimal numbers. You must also be able to convert decimal numbers to binary numbers.

In Fig. 1-6, you can see that binary 10001100100.11 is equal to $1024 + 64 + 32 + 4 + 0.5 + 0.25 = 1124.75$. Now let's see how the 1124.75 is converted to the binary notation.

Step 1 (Fig. 1-6) records the number to be converted. Step 2 is to write down the largest whole power of 2 (in decimal form)

POWERS OF 2	2^{11}	2^{10}	2^9	2^8	2^7	2^6	2^5	2^4	2^3	2^2	2^1	2^0	•	2^{-1}	2^{-2}
DECIMAL VALUE	2048	1024	512	256	128	64	32	16	8	4	2	1	•	0.5	0.25
BINARY FOR 1124.75	0	1	0	0	0	1	1	0	0	1	0	0	•	1	1

STEP		NOTES	OPERATION
1	1124.75	Number To Be Converted	
2	− 1024.00	Largest Whole Power of 2 That Can Be Subtracted − − − −	Place a 1 Under 1024
3	100.75	Remainder	
4	− 64.00	Largest Whole Power of 2 That Can Be Subtracted − − − −	Place a 1 Under 64
5	36.75	Remainder	
6	− 32.00	Largest Whole Power of 2 That Can Be Subtracted − − − −	Place a 1 Under 32
7	4.75	Remainder	
8	− 4.00	Largest Whole Power of 2 That Can Be Subtracted − − − −	Place a 1 Under 4
9	0.75	Remainder	
10	− 0.50	Largest Whole Power of 2 That Can Be Subtracted − − − −	Place a 1 Under 0.5
11	0.25	Remainder	
12	− 0.25	Largest Whole Power of 2 That Can Be Subtracted − − − −	Place a 1 Under 0.25
13	0.00	Remainder	

Fig. 1-6. Steps in converting decimal 1124.75 to its binary equivalent.

that can be subtracted from the number in step 1. This is 1024, so you place a 1 under the decimal value of 1024 in the bottom row of Fig. 1-6. Step 3 is to find the remainder, and Step 4 is to find the largest whole power of 2 that can be subtracted from the 100.75 of step 3. So you place a 1 under 64, and so on through all remaining steps. Then you write a 0 under all values not used in the above procedure. Thus you find that decimal 1124.75 = 10001100100.11. When you memorize the powers of 2 and their decimal equivalents, and get a lot of practice (which is required in *any* math), you will be able to write a binary notation of a decimal number without actually drawing up a table.

	DECIMAL NUMBER	÷	BASE 2	=	RESULT	REMAINDER (BINARY)	
	54	÷	2	=	27	0	LSB
Fig. 1-7. Conversion of decimal	27	÷	2	=	13	1	
54 to binary form.	13	÷	2	=	6	1	
	6	÷	2	=	3	0	
	3	÷	2	=	1	1	READ UP
	1	÷	2	=	0	1	MSB

= BINARY 1 1 0 1 1 0

An alternate form of converting decimal to binary is illustrated by Fig. 1-7. In this example, 54 is converted to binary by repeatedly dividing by base 2 and recording the remainders. The lsb is at the top with the msb at the bottom, so read up to get the equivalent binary number. Double check the binary by noting that 110110 is 32+16+4+2 = 54 decimal. Again note that you will be able to jot the resulting binary down mentally with a little practice. Remember to reverse the result to get the msb first.

1-7. FROM BINARY NUMBERS TO BOOLEAN NOTATION

Boolean algebra concerns itself with logical operations, whether these operations are mathematical or statements of results to be obtained.

Variables A, B, C, and D each must have either a 0 or 1 value. If each has the value of 1, then:

Binary 01 = $\bar{A}B$ Boolean
10 = $A\bar{B}$ Boolean

$$101 = A\bar{B}C \text{ Boolean}$$
$$1010 = A\bar{B}C\bar{D} \text{ Boolean}$$

and so on. For example, Table 1-7 lists the basic binary addition function. This function is termed a *half-adder* since there is provision for a carry output but no carry input. Note that the sum column has an output (logical 1) only when the inputs are of opposite polarity ($A\bar{B}$ or $\bar{A}B$). When the inputs are of like polarity (either $\bar{A}\bar{B}$ or AB), no output (logical 0) occurs. The carry output exists only when two ones exist (AB) to meet the requirements of binary addition.

Table 1-7. Basic Binary Addition Function

A	Plus	B	=	Sum	Out	Carry Out	Boolean Notation
0		0		0		0	$\bar{A}\bar{B}$
0		1		1		0	$\bar{A}B$
1		0		1		0	$A\bar{B}$
1		1		0		1	AB

Table 1-8. Full Adder in Boolean Form

A	B	C	Boolean Sum ($f = 1$)	Boolean New Carry (C_n)
0	0	0	0	0
0	0	1	$\bar{A}\bar{B}C$	0
0	1	0	$\bar{A}B\bar{C}$	0
0	1	1	0	$\bar{A}BC$
1	0	0	$A\bar{B}\bar{C}$	0
1	0	1	0	$A\bar{B}C$
1	1	0	0	$AB\bar{C}$
1	1	1	ABC	ABC

C = Previous carry
C_n = New carry

A full binary adder must handle a carry input as well as provide a carry output. (Review Table 1-4.) Now rewrite the binary addition function of Table 1-4 showing ($f=1$ for sum and $f=1$ for carry in Boolean notation as in Table 1-8, where C is the previous carry (carry input), and C_n is the new carry (carry output). Then from Table 1-8 note that:

$$Sum = \overline{A}\overline{B}C + \overline{A}B\overline{C} + A\overline{B}\overline{C} + ABC$$

$$C_n = \overline{A}BC + A\overline{B}C + AB\overline{C} + ABC$$

Be sure you can correlate the Boolean notation of Table 1-8 with the binary notation of Table 1-4.

You will learn in future chapters how these relationships can be greatly simplified to reduce hardware requirements and cost by the use of Boolean algebra. In fact, this is the primary reason for mastering the subject.

EXERCISES

1-1. A logic system employs two voltages: (a) 0 volts and (b) −4 volts. For positive logic, which is logical 1 and which is logical 0?

1-2. Name the three basic operations in Boolean algebra.

1-3. What does the operator NOT mean in Boolean notation?

1-4. Read aloud the Boolean output notation $AB + \overline{A}\overline{B}$.

1-5. For Question 1-4, if you have $\overline{A}B + A\overline{B}$ as inputs, what would you expect?

1-6. Convert the following binary numbers to their decimal (base 10) equivalents:
   (a) 101001.01
   (b) 1011.1111
   (c) 11101010.101
   (d) 0001.0001

1-7. Find the binary sum and give the decimal equivalent of:
   (a) 1000 + 1001
   (b) 1101.1 + 1000.101
   (c) 0000.1110 + 0001.1110
   (d) 0000.11111 + 0011.111

1-8. Find the difference of the following by two methods: straight arithmetical and by using complements.
   (a) 1110 − 0110
   (b) 101001 − 001010

CHAPTER 2

Symbolic Logic and Truth Tables

The symbolic form of Boolean algebra consists of symbols representing Boolean operations, and by alphabetic letters representing the relationships.

2-1. LOGIC STATEMENTS

Let's make a two-part statement:

A. "The object is a car."

AND

B. "The car is blue."

A statement is either true (T) or false (F). Often the result or functions of *combined* statement A AND B (AB) is denoted by a function (f). The connective (AND in this case) is also called the Boolean *operator*.

So if you say, "The object is a car AND the car is blue," the function of this combined statement can be true only if *both* A AND B are true. If either or both are false, the combined statement (f) is false. The truth table for this operation is shown in Table 2-1. Reason as follows:

Table 2-1. Truth Table for AND Operation

Row	Inputs A	B	Output f = AB
1	F	F	F
2	F	T	F
3	T	F	F
4	T	T	T

Row 1. If A is false and B is false, the combined function (f) is false.

Row 2. If A is false and B is true, the combined function is false.

Row 3. If A is true and B is false, the combined function is false.

Row 4. If A is true and B is true, the combined function is true.

> NOTE: Get very familiar with "truth tables." Most component specifications are given in truth table form. Technicians troubleshoot components and entire systems by comparing signals or levels of truth table values. Practice using the truth table form until you can draw up a truth table to meet any specific requirement. The technique will be emphasized in this chapter.

Any AND operation requires that all inputs be coincidently high (logical 1) to obtain an output. For a two-part statement, A AND B (AB) must both be true (logical 1) to obtain a logical 1 output. For a four-part statement, ABCD, each variable must all be true to obtain an output; and so on.

Thus if you say "It is snowing *and* two plus two equals four," the statement is true only if it is snowing. The statement "It is snowing and two plus two equals five" is false no matter what the state of the weather. Stated in Boolean arithmetic form:

$\bar{A}\bar{B}$ = false

$\bar{A}B$ = false

$A\bar{B}$ = false

AB = true

Now let's make another two-part statement:

A. "Jack will be there."

OR

B. "Jill will be there."

There are two types of *ors* found in the study of logic:

1. *Inclusive or:* either one, or both.
2. *Exclusive or:* either one, but not both.

Thus for the *inclusive* OR, if either Jack OR Jill or both Jack and Jill are present, the statement is true (Table 2-2). If the *exclusive* OR is inferred, if either Jack or Jill is there, the statement is true. If both are there, the statement is false (Table 2-3). For both OR operations, if *neither* Jack nor Jill is present, the statement is false. We will study the two basic types of OR operation as the chapter progresses.

Table 2-2. Truth Table for Inclusive-OR Operation

Row	Inputs A B	Output f = A + B
1	F F	F
2	F T	T
3	T F	T
4	T T	T

Note carefully that in the construction of a truth table, the Boolean variables A, B, etc., are always listed in their true form at the head of the column, and that the actual value of each variable can be either true (T) or false (F). Thus all possible combinations of values are listed in a truth table; the left part of the table contains all combinations of values of the variables used in the expression, and the right part contains the value of the expression for each combination of values listed on the left

Table 2-3. Truth Table for Exclusive-OR Operation

Row	Inputs A B	Output A ⊕ B
1	F F	F
2	F T	T
3	T F	T
4	T T	F

side. A truth table tells what the output level should be for every possible combination of input levels. Every possible combination of inputs is found by simple binary addition. Where T = 1 and F = 0, for a two-input device:

Row 1. 00

Row 2. $\dfrac{1}{01}$ add 1

Row 3. $\dfrac{1}{10}$ add 1

Row 4. $\dfrac{1}{11}$ add 1

2-2. THE CONNECTIVE (OPERATOR) AND

Basic logic is founded on either a true or false condition with no "gray area" in between. A true or false "AND" condition can be very simply represented by a closed or open switch, as shown by Fig. 2-1A.

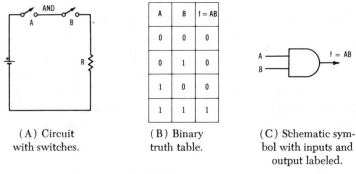

A	B	f = AB
0	0	0
0	1	0
1	0	0
1	1	1

(A) Circuit with switches.   (B) Binary truth table.   (C) Schematic symbol with inputs and output labeled.

Fig. 2-1. AND circuit.

If either A or B is open, no current will flow in R. If A AND B are closed, current will flow in R.

Let a closed switch = true = 1.

Let an open switch = false = 0.

Then for the two-part "AND" statement of Section 2-1, Fig. 2-1B shows the truth table in *binary* notation.

The schematic symbol for an AND circuit is shown by Fig. 2-1C, which says that an output (f=1) will occur *only* if both A AND B are 1s. When you see this symbol, you should recognize the AND function.

Fig. 2-2 is a summary of the AND function. In row 1, switch A must be a two-contact series arrangement. Therefore the function depends entirely on the "A" mode. It is read "A AND A = A."

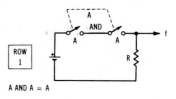

A AND A = A

Fig. 2-2. Summary of AND circuit.

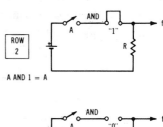

A AND 1 = A

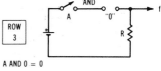

A AND 0 = 0

In row 2, since the "1" is a closed circuit, the function depends entirely on the "A" mode. It is read "A AND 1 = A."

In row 3, the "0" is an open circuit; therefore the AND function will always be 0 regardless of the "A" mode. The function is read "A AND 0 = 0."

An open or closed switch can be represented by Boolean notation. If A =1, $\bar{A}$ = 0. The bar over the A ($\bar{A}$) represents inverted A. If B = 1, $\bar{B}$ = 0. The bar over the B ($\bar{B}$) represents inverted B.

Table 2-4 includes the Boolean notation in the AND truth table. When switch A is closed (binary 1), the Boolean notation is A. When switch A is open (binary 0), the Boolean notation is $\bar{A}$ (read "A NOT"). The same notation prevails for switch B.

Table 2-4. AND Truth Table With Boolean Notation

Binary		Boolean	f = AB	Comments
A	B			
0	0	$\bar{A}\bar{B}$	0	Reads "A NOT AND B NOT = 0"
0	1	$\bar{A}B$	0	Reads "A NOT AND B = 0"
1	0	$A\bar{B}$	0	Reads "A AND B NOT = 0"
1	1	AB	1	Reads "A AND B = 1"

NOTE: $\bar{A}$ can also be read as NOT A. Similarly, $\bar{B}$ can be read as NOT B. However, there are many times when the A NOT and B NOT reading is preferable. An example of this will be shown in Section 2-5.

An AND logic circuit is normally called a gate. Fig. 2-3A shows a pulse application in which different width pulses are applied to inputs A and B. An output can occur from the AND gate only when both inputs are high (coincident). Therefore the input on A is "gated" at times when the input on B is coincidently high.

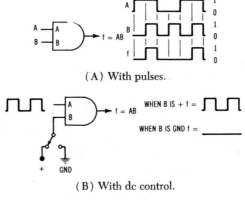

(A) With pulses.

(B) With dc control.

Fig. 2-3. Operation of an AND gate.

Fig. 2-3B shows an AND gate where input B is dc-switched from plus (high) to ground (low). Thus the A input is "gated" to the output only when input B is high (positive voltage). In general, you should understand that the logic symbol is used either for a logical operation, or as a gated operation.

2-3. THE CONNECTIVE OR

Another basic logic function is the OR circuit. There are two types of OR function:

1. The inclusive OR, either one or both (and/or).
2. The exclusive OR, either one but not both (either/or but not both).

The Inclusive OR

Let's make two statements:

A. Money will be issued

      OR

B. A universal credit card will be issued.

Symbols: A OR B = A + B

        f = function of combined statement A+B (A OR B)

If you say, "Money will be issued OR a universal credit card will be issued," the combined statement will be true if A is true, OR if B is true, OR *both* A *and* B are true (one or both). This is the "inclusive OR" function. The schematic symbol is shown in Fig. 2-4A, and the inclusive OR truth table by Fig. 2-4B.

Inclusive OR logic can be very simply represented by open or closed switches as in Fig. 2-4C. If switch A OR switch B is closed, current will flow in R. If both A and B are closed, current will flow in R. If A and B are both open, no current will flow. As

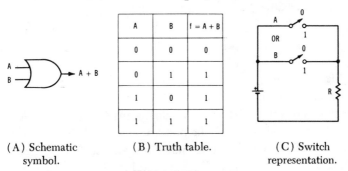

A	B	f = A + B
0	0	0
0	1	1
1	0	1
1	1	1

(A) Schematic symbol. (B) Truth table. (C) Switch representation.

Fig. 2-4. Inclusive OR function.

before, a closed switch can represent 1, and an open switch 0. Observe that whereas the AND function must be represented as a series circuit, the OR function must be represented as a parallel circuit.

Note carefully that the inclusive OR circuit can be either variable true or both true. The usual definition is that $f = A+B$. The complete definition is actually that $f = A+B+AB$. This reads:

$$f = A \text{ OR } B \text{ OR } A \text{ AND } B.$$

The combined statement (f) can be true if one is true and the other is false. Thus a true statement and a false statement connected by the connective OR is a true statement (if A and/or B is true). If *neither the* money nor the credit card is issued (both A and B are false or 0), then the statement is false.

The Exclusive OR

Now make two statements:

A. You will go to Chicago on Friday

     OR ◄——— connective

B. You will go to San Francisco on Friday.

This combined statement implies that either A OR B can be true, but not both. Thus the "exclusive OR" function is required.

The symbol for the exclusive OR connective is ⊕ (either/or but not both). The exclusive OR symbol is shown in Fig. 2-5A. The exclusive OR truth table is presented in Fig. 2-5B. Note that to produce an output (1), the inputs must be opposite. Like inputs (either 0 or 1) result in no output (0). Thus $\bar{A}B$ OR $A\bar{B}$ (read A NOT AND B OR A AND B NOT) is required for an output. Thus $A \oplus B = \bar{A}B + A\bar{B}$.

Fig. 2-5C shows a discrete circuit illustrating the exclusive OR function. Although IC (integrated circuit) chips are normally used, this discrete circuit allows illustrating the basic action.

If both A and B are of like polarity (0s or 1s), both transistors have zero-biased base-emitter junctions, and neither can conduct. Thus the output is the full negative value of $-V_{CC}$. Now assume A is 0 (ground) and B is +5 volts ($\bar{A}B$). Thus the base of Q_2

A	B	f = A ⊕ B
0	0	0
0	1	1
1	0	1
1	1	0

(A) Schematic symbol. (B) Truth table.

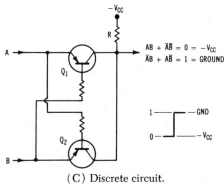

(C) Discrete circuit.

Fig. 2-5. Exclusive OR function.

is grounded and the emitter is at +5 volts, causing it to conduct. This sends the output to essentially ground, creating the "high" (ground) condition, producing a "1" output. If A is +5 volts and B is grounded (0) (the condition for $A\bar{B}$), then Q_1 conducts, also producing a 1 output. Thus the condition for $\bar{A}B + A\bar{B} = 1$ is satisfied.

2-4. THE NOT OPERATION

Up to this point, every *statement* (or *proposition*) has been positive in content. Thus when you say, "The oscillator output level is normal *and* the frequency is within tolerance," you are making a positive statement that is either true or not true (false).

Every statement has an opposite. For example:

A = the oscillator output level is normal.
$\bar{A}$ = the oscillator output level is NOT normal.

35

B = the frequency is within tolerance.

$\bar{B}$ = the frequency is NOT within tolerance.

The bar over the A ($\bar{A}$) indicates the negative, or inverted form, of A. Similarly, the bar over the B ($\bar{B}$) indicates the negative, or inverted form of B.

If A = 1, $\bar{A}$ = 0. (If A is high, $\bar{A}$ is low.)
If A = 0, $\bar{A}$ = 1. (If A is low, $\bar{A}$ is high.)
If B = 1, $\bar{B}$ = 0. (If B is high, $\bar{B}$ is low.)
If B = 0, $\bar{B}$ = 1. (If B is low, $\bar{B}$ is high.)

Sometimes a "straight inverting" stage (NOT gate) is used for logic inversion. The symbol is shown by Fig. 2-6A. The circle on the output means a "1" input (high level) results in a "0" (low level) output. It is termed a high-level activated stage. If the circle appears at the input (Fig. 2-6B), a 0 (low) input results in a 1 (high) output (low-level activated). Either type is inverting. The circles at input or output are important only to clearly indicate the significant function. A circle always indicates signal inversion at that point.

(A) High-level activated.   (B) Low-level activated.

Fig. 2-6. Schematic symbols for NOT circuits.

A given character passed through one inverting stage gives the reversed polarity original character. A given character passed through two consecutive inverting stages gives the original character polarity (see Fig. 2-7A).

This is the same as passing any signal through two common-cathode (tube) stages or common-emitter (transistor) stages.

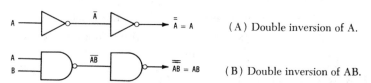

(A) Double inversion of A.

(B) Double inversion of AB.

Fig. 2-7. Examples of double inversion.

The original polarity results, i.e., $\bar{\bar{A}} = A$. Note that the same holds true for $\bar{\bar{AB}} = AB$ as in Fig. 2-7B.

The above simply illustrates a statement that contains a double negative. If you say "It is not not raining" this means the same as "It is raining." In equation form:

(It is NOT NOT raining) = (It is raining)

2-5. VARIATIONS OF THE CONNECTIVE AND

A NOT AND circuit is termed a NAND gate. The schematic symbol is shown by Fig. 2-8A. The term $\overline{AB}$ is read "NOT A AND B." The small circle on the output indicates polarity reversal. This says that if A and B are both true (1), the output ($\overline{AB}$) is false (0). It is the "inverted" AND function.

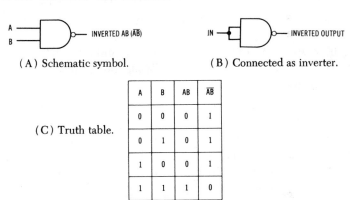

(A) Schematic symbol.   (B) Connected as inverter.

(C) Truth table.

A	B	AB	$\overline{AB}$
0	0	0	1
0	1	0	1
1	0	0	1
1	1	1	0

Fig. 2-8. The NAND gate.

Fig. 2-8B shows how an ordinary NAND gate can be connected as a straight inverter. Since both inputs are tied together, a "1" on the input will result in a "0" (inverted) output, and vice versa.

Fig. 2-8C shows the truth table for the AND function and NAND function. Note that if one input of a NAND gate is 0, the output will always be 1. The output is the reversal of AB.

The *inhibitor* gate (Fig. 2-9) produces an output only when the inputs represent $A\bar{B}$ (A AND B NOT) or $\bar{A}B$ (A NOT AND B). Fig. 2-9A is the schematic symbol for $f = A\bar{B}$. The $\bar{B}$ may be obtained from a straight inverter stage preceding the B input.

The small circle at the B input represents inversion. A 1 input at B becomes a 0 input to the AND gate. A 0 input at B becomes a 1 input to the AND gate. Thus f=A$\bar{\text{B}}$ (read A AND B NOT). For a 1 to occur at the output, A must be 1 and B must be 0. Obviously if the circle is on the A input, f = $\bar{\text{A}}$B (read as A NOT AND B), and conditions are reversed.

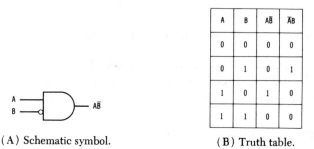

A	B	A$\bar{\text{B}}$	$\bar{\text{A}}$B
0	0	0	0
0	1	0	1
1	0	1	0
1	1	0	0

(A) Schematic symbol.   (B) Truth table.

Fig. 2-9. Inhibitor gate.

Fig. 2-9B shows the truth table for A$\bar{\text{B}}$ and $\bar{\text{A}}$B. Note that for f=A$\bar{\text{B}}$, the output is inhibited (0) for all conditions except when A=1 AND B=0. For f=$\bar{\text{A}}$B, output is inhibited except when A=0 AND B=1. In either case, inputs must be of opposite character for an output to occur.

It was stated in Section 2-2 that $\bar{\text{A}}$ can be read either "NOT A" or "A NOT", and also that $\bar{\text{B}}$ can be read either "NOT B" or "B NOT." The "A NOT" and "B NOT" readings are preferred as can be seen by the following example.

Assume you have the condition of $\bar{\text{A}}$B. If you read this as "NOT A AND B," you are stating the condition of $\overline{\text{AB}}$ (read as "NOT A

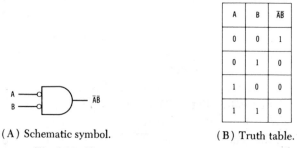

A	B	$\bar{\text{A}}\bar{\text{B}}$
0	0	1
0	1	0
1	0	0
1	1	0

(A) Schematic symbol.   (B) Truth table.

Fig. 2-10. Characteristics of A NOT AND B NOT gate.

AND B"). But note the decided difference between Figs. 2-8C and 2-9B. Thus you should use the term "A NOT AND B" for f = ĀB.

When small circles (indicating inversion) are at both inputs of an AND gate as in Fig. 2-10A, f = ĀB̄ reads f = A NOT AND B NOT. Fig. 2-10B shows the truth table for ĀB̄. The output function is a 1 when (and only when) both inputs are 0. Note carefully that ĀB̄ (A NOT AND B NOT) is different from $\overline{AB}$ (NOT A AND B). Review Fig. 2-8C.

2-6. VARIATIONS OF THE CONNECTIVE OR

Inversion is used with OR gates as well as with AND gates. The term for a NOT OR gate is a NOR gate. The schematic symbol is that of Fig. 2-11A. The term $\overline{A+B}$ is read as NOT A OR B. The small circle on the output indicates the polarity inversion. The circuit is a reversed-polarity OR gate. Only when both inputs are low (0) is the output high (1). Fig. 2-11B shows the truth table of A+B and $\overline{A+B}$ (OR and NOR gates).

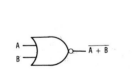

A	B	A+B	$\overline{A+B}$
0	0	0	1
0	1	1	0
1	0	1	0
1	1	1	0

(A) Schematic symbol for NOR gate.

(B) Truth table for OR and NOR.

Fig. 2-11. The NOR gate.

Three-state (sometimes termed *tri-level*) logic devices have an extra input termed an *enable-disable* gating input. When logic is *enabled*, the output is either at a logic 0 or logic 1, but not both. When *disabled*, the output is disconnected from the rest of the circuit. Fig. 2-12A illustrates the extra input that *enables* the gate with a logic 0. Fig. 2-12B shows the extra input that *enables* the circuit with a logic 1.

Three-state or three-level logic devices are not limited to NOR gates. In practice, AND, NAND, OR, and NOR gates as well as straight inverting and noninverting stages are available with enable/disable inputs. (The name Tri-State is a registered trademark of National Semiconductor Corp.)

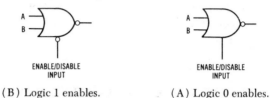

ENABLE/DISABLE INPUT   ENABLE/DISABLE INPUT

(B) Logic 1 enables.   (A) Logic 0 enables.

Fig. 2-12. Symbols for three-state NOR gates.

An output can be obtained under all conditions except when A=0 and B=1 by the circuit of Fig. 2-13A. The function $f=A+\bar{B}$ is read "A OR B NOT." The small circle on the B input indicates inversion. A 1 at the B input becomes a 0 at the OR gate input. A 0 at the B input becomes a 1 at the OR input. If the circle were at the A input, then $f=\bar{A}+B$ (read as A NOT OR B). Fig. 2-13B is the truth table for $A+\bar{B}$ and $\bar{A}+B$. For clarity, the inverted values of A and B are included in Fig. 2-13B.

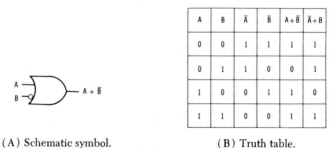

A	B	$\bar{A}$	$\bar{B}$	$A+\bar{B}$	$\bar{A}+B$
0	0	1	1	1	1
0	1	1	0	0	1
1	0	0	1	1	0
1	1	0	0	1	1

(A) Schematic symbol.   (B) Truth table.

Fig. 2-13. Gate for A OR B NOT function.

When small circles (indicating inversion) are on both inputs of an OR gate as in Fig. 2-14A, then $f=\bar{A}+\bar{B}$, which is read as "A NOT OR B NOT." Only when both inputs are high (1) is the output low (0). Fig. 2-14B is the truth table for $\bar{A}+\bar{B}$. Compare this truth table to Fig. 2-11B. Note that $\bar{A}+\bar{B}$ is *not* the same as $\overline{A+B}$.

When an OR gate is specified, the inclusive OR gate is meant.

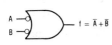

A	B	$\overline{A}+\overline{B}$
0	0	1
0	1	1
1	0	1
1	1	0

(A) Schematic symbol.　　　　(B) Truth table.

Fig. 2-14. Gate for A NOT OR B NOT.

The schematic symbol for the inverted exclusive OR (exclusive NOR) is shown in Fig. 2-15A. The truth table for exclusive NOR is given in Fig. 2-15B. Note that $\overline{A \oplus B} = \overline{A}\overline{B} + AB$, which reads as "A NOT AND B NOT OR A AND B."

A discrete circuit that produces $\overline{A \oplus B}$ is shown in Fig. 2-15C. When both A and B are alike (either 0 or 1) Q_1 and Q_2 have zero-biased base-emitter junctions, and form open switches. With

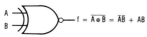

A	B	$\overline{A}\overline{B} + AB$
0	0	1
0	1	0
1	0	0
1	1	1

(A) Schematic symbol.　　　　(B) Truth table.

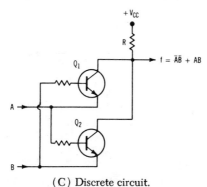

(C) Discrete circuit.

Fig. 2-15. Exclusive NOR gate.

41

no current through R, $f=V_{CC}=high=1$. When A and B differ $(0, 1)$ the opposite logic levels cause either Q_1 or Q_2 to conduct. The resultant current in R sends the output to essentially ground (low level, or 0). Note that this is just the inverse of the exclusive OR gate, which requires *unlike* input polarities to produce an output (Fig. 2-5C). Note also that Figs. 2-5C and 2-14C are simply inverse circuits; the type of transistors (npn or pnp) and voltage polarity are reversed.

2-7. LOGIC HARDWARE

The "black boxes" we have discussed that perform Boolean functions are normally integrated circuits (ICs) that have defined

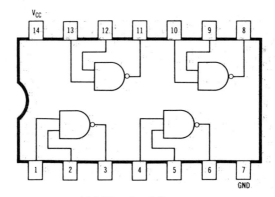

(A) Functional diagram.

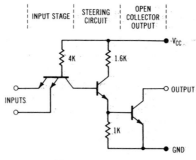

(B) Circuit of one gate.

Fig. 2-16. Quad two-input NAND gate.

characteristics. Such an IC may contain a single gate, or many circuit functions equivalent to hundreds of transistors and electronic components. Fig. 2-16 represents one type of IC (14-pin) that is approximately ¾ inch long by ¼ inch wide and ¼ inch deep, including pins. This particular IC is a quad two-input NAND gate. Note that a single V_{CC} pin (usually +5 volts) and one ground pin are provided for the operation of all four gates. Power supply connections are never shown in normal logic schematics and must be assumed by the reader.

EXERCISES

2-1. Read aloud the expression: (a) AB, (b) $\overline{AB}$, (c) $\overline{A}\overline{B}$, (d) A+B, (e) $\overline{A+B}$, (f) A ⊕ B.

2-2. Give another equivalent symbolic expression for A ⊕ B.

2-3. Read aloud the expression $\overline{A}B+A\overline{B}$.

2-4. Prepare the truth table for the circuit of Fig. 2-17.

Fig. 2-17. Logic circuit for
Exercise 2-4.

Table 2-5. Truth Table for A + B̄

Inputs		Output
A	B	A + B̄
0	0	
0	1	
1	0	
1	1	

2-5. Write the binary relationship A=1 OR B=0 OR C=0 OR D=1 in Boolean form.

2-6. Complete Table 2-5.

2-7. Complete Table 2-6.

2-8. Complete Table 2-7.

Table 2-6. Truth Table for ABC̄

| | Inputs | | Output |
A	B	C	ABC̄
0	0	0	
0	0	1	
0	1	0	
0	1	1	
1	0	0	
1	0	1	
1	1	0	
1	1	1	

Table 2-7. Truth Table for A + B + C̄

| | Inputs | | Output |
A	B	C	A+B+C̄
0	0	0	
0	0	1	
0	1	0	
0	1	1	
1	0	0	
1	0	1	
1	1	0	
1	1	1	

Algebraic Operations for Circuit Simplification

An important item in the study of logic gates for computers is the "equivalence" of certain functions.

3-1. EQUIVALENCE (DUALITY) OF LOGIC GATES

Table 3-1 presents the comparative truth table for many of the logic circuits covered thus far. You can pick out any duplication of columns of truth values in this truth table, and consider the functions as equivalent. For example, note that $\overline{AB}$ and $\bar{A}+\bar{B}$ have the same truth values (function). Therefore $\overline{AB}$ is equivalent

Table 3-1. Truth Table Showing Dualities

A	B	AB	$\overline{AB}$	$\overline{\bar{A}\bar{B}}$	A + B	$\overline{A+B}$	$\bar{A}+\bar{B}$
0	0	0	1	1	0	1	1
0	1	0	1	0	1	0	1
1	0	0	1	0	1	0	1
1	1	1	0	0	1	0	0

ROW	AND (NAND)	TRUTH TABLE			OR (NOR)
1		A	B	f	
		0	0	0	
		0	1	0	
		1	0	0	
		1	1	1	
2		A	B	f	
		0	0	0	
		0	1	1	
		1	0	0	
		1	1	0	
3		A	B	f	
		0	0	0	
		0	1	0	
		1	0	1	
		1	1	0	
4		A	B	f	
		0	0	1	
		0	1	0	
		1	0	0	
		1	1	0	
5		A	B	f	
		0	0	0	
		0	1	1	
		1	0	1	
		1	1	1	
6		A	B	f	
		0	0	1	
		0	1	1	
		1	0	0	
		1	1	1	
7		A	B	f	
		0	0	1	
		0	1	0	
		1	0	1	
		1	1	1	
8		A	B	f	
		0	0	1	
		0	1	1	
		1	0	1	
		1	1	0	

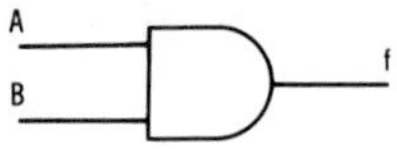

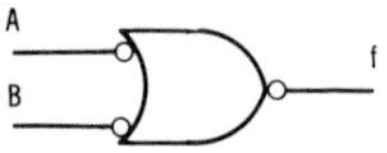

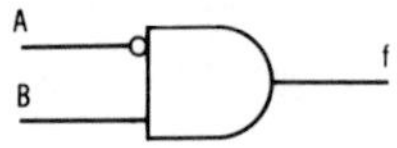

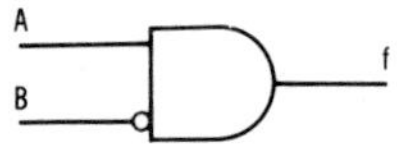

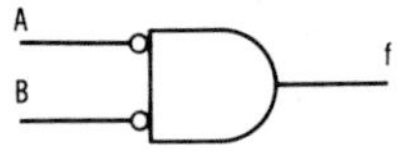

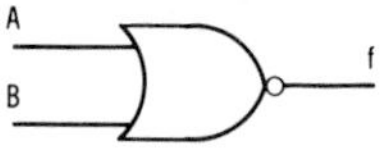

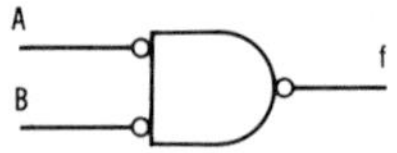

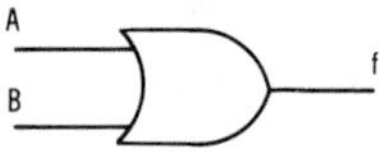

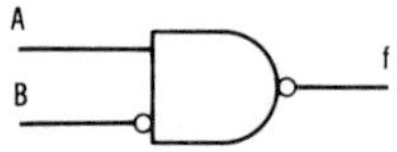

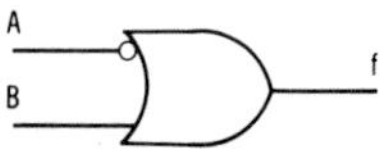

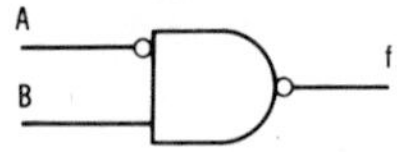

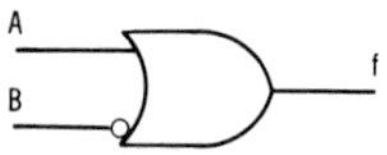

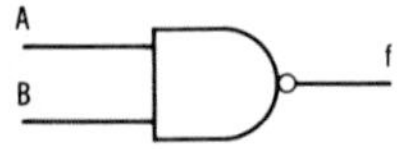

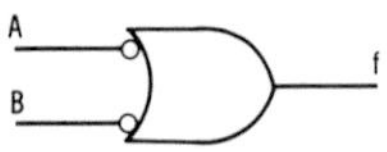

Fig. 3-1. Equivalent logic gates.

to $\bar{A}+\bar{B}$ (read as NOT A AND B is *equivalent* to A NOT OR B NOT). Thus a duality exists for $\overline{AB}$ and $\bar{A}+\bar{B}$.

Also note that $\bar{A}\bar{B}$ has the same truth values, or function, as $\overline{A+B}$. Therefore $\bar{A}\bar{B}$ is equivalent to $\overline{A+B}$ (read A NOT AND B NOT is *equivalent* to NOT A OR B). Thus a duality exists for $\bar{A}\bar{B}$ and $\overline{A+B}$.

Fig. 3-1 reviews the truth tables and symbols for logic gates, and shows the dual nature of such gates. Study this figure and be sure you understand the duality of the symbols shown.

3-2. LOGIC NETWORKS

Logic networks (combined AND, NAND, OR, and NOR gates) are used to satisfy Boolean relationships. In the design of logic systems, the functions are first reduced to Boolean form. Fig. 3-2 illustrates an example of a logic network to perform the exclusive OR function. Note that $f = \bar{A}B + A\bar{B} = A \oplus B = $ exclusive OR.

Although multiple components can be and sometimes are used to perform this function, exclusive OR gates (ICs) are available with the function built in on one chip. Examples are the Motorola quad exclusive-OR Type MC 14507CL, or MC 771P, and the Signetics Type SN 7486.

Table 3-2 lists all the binary and Boolean relationships, with rule numbers for future reference.

Fig. 3-3 proves Rule 24, that $A(A+B) = A$, by switch contacts. Note that the output function depends entirely upon the 0 or 1 position of A, regardless of the B mode.

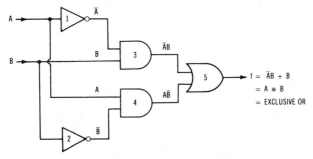

Fig. 3-2. Example of logic network.

Table 3-2. Binary and Boolean Relationships

Rule No.	Relation	Laws	
1	$0 + 0 = 0$		
2	$(0)(0) = 0$		
3	$1 + 1 = 1$		
4	$(1)(1) = 1$	Binary	
5	$(0)(1) = 0$		
6	$0 + 1 = 1$		
7	$\bar{0} = 1$		
8	$\bar{1} = 0$		
9	$A + A = A$		
10	$(A)(A) = A$		
11	$(\bar{A})(A) = 0$		
12	$\bar{A} + A = 1$	Boolean-Binary	
13	$0 + A = A$		
14	$(0)(A) = 0$		
15	$1 + A = 1$		
16	$(1)(A) = A$		
17	$A + B = B + A$	Commutative	Three basic algebraic laws applicable to Boolean expression
18	$(A)(B) = (B)(A)$		
19	$(A + B) + C = A + (B + C)$	Associative	
20	$(AB)C = A(BC)$		
21	$A(B + C) = AB + AC$	Distributive	
22	$A + AB = A$		
23	$A + \bar{A}B = A + B$	Simplification Rules and DeMorgan's Theorem	
24	$A(A + B) = A$		
25	$\overline{AB} = \bar{A} + \bar{B}$		
26	$\overline{A + B} = \bar{A}\bar{B}$		

NOTE: Multiple letters within parentheses must be considered as a group. Thus, for $A(A + B)$ the group $(A + B)$ is either a 0 or a 1. The relation $A(A + B)$ is read "A and the quantity A OR B." This network would be an AND/OR combination.

To prove that $A(A+B) = A$ by the algebraic method:

$$A(A+B) = AA+AB \qquad \text{Rule 21}$$
$$AA+AB = A+AB \qquad \text{Rule 10}$$
$$A+AB = (1)(A) + AB \qquad \text{Rule 16}$$

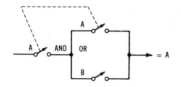

Fig. 3-3. Use of switch contacts to prove that $A(A + B) = A$.

$$(1)(A) + AB = A(1+B) \qquad \text{Rule 21}$$
$$A(1+B) = (A)(1) \qquad \text{Rule 15}$$
$$(A)(1) = A \qquad \text{Rule 16}$$
$$\text{Therefore } A(A+B) = A \qquad \text{(proof of Rule 24)}$$

Table 3-3 proves that $A(A+B) = A$ by the truth table method.

Table 3-3. Proof That A(A+B) = A by Truth Table

Row	Column			
	1	2	3	4
1	A	B	A + B	A(A + B)
2	0	0	0	0
3	0	1	1	0
4	1	0	1	1
5	1	1	1	1

Steps in truth table form of simplification or demonstration are:

1. List the variables and all combinations called for in the expression to be simplified (row 1, columns 1–4 in this example).
2. Put down all possible binary values of the elementary letters. (A and B in this example, rows 2 through 5, columns 1 and 2).
3. Determine the values of the elementary combinations called for in the expression to be simplified (A+B in this example, rows 2 through 5).
4. List the values of the combinations of combinations (rows 2 through 5, column 4, in this example).

Note that columns A and A(A+B) are identical. Since the A+B part of A(A+B) is superfluous, the simplest term is A. This means that $A(A+B) = A$.

Another example is to simplify $A\bar{B}\bar{C} + ABC + \bar{A}BC$ algebraically:

$$A\bar{B}\bar{C} + ABC + \bar{A}BC = A\bar{B}\bar{C} + BC(A+\bar{A}) \qquad \text{Rule 21}$$
$$= A\bar{B}\bar{C} + BC \qquad \text{Rule 12}$$

It is imperative that the simplest relationship possible be used in logic design. Sometimes it is not readily evident as to when final simplification has been reached algebraically. *Karnaugh maps* (sometimes termed *Veitch diagrams*) are often used. This subject is covered in Chapter 5.

Also, the chart form of simplification can be used as in Chart 3-1. This example is for three variables. As many variables as necessary can be handled in chart form. Note that all possible combinations of ABC must be used. The procedure for simplifying $A\bar{B}\bar{C} + ABC + \bar{A}BC$ follows:

Chart 3-1. Example of Chart Form of Simplification

Row	Column 1 A	2 B	3 C	4 AB	5 AC	6 BC	7 ABC
1	$\bar{A}$	$\bar{B}$	$\bar{C}$	$\bar{A}\bar{B}$	$\bar{A}\bar{C}$	$\bar{B}\bar{C}$	$\bar{A}\bar{B}\bar{C}$
2	$\bar{A}$	$\bar{B}$	C	$\bar{A}\bar{B}$	$\bar{A}C$	$\bar{B}C$	$\bar{A}\bar{B}C$
3	$\bar{A}$	B	$\bar{C}$	$\bar{A}B$	$\bar{A}\bar{C}$	$B\bar{C}$	$\bar{A}B\bar{C}$
4	$\bar{A}$	B	C	$\bar{A}B$	$\bar{A}C$	$\boxed{BC}$	$\bar{A}BC$
5	A	$\bar{B}$	$\bar{C}$	$A\bar{B}$	$A\bar{C}$	$\bar{B}\bar{C}$	$\boxed{A\bar{B}\bar{C}}$
6	A	$\bar{B}$	C	$A\bar{B}$	AC	$\bar{B}C$	$A\bar{B}C$
7	A	B	$\bar{C}$	AB	$A\bar{C}$	$B\bar{C}$	$AB\bar{C}$
8	A	B	C	AB	AC	$\boxed{BC}$	ABC

Simplify: $A\bar{B}\bar{C} + ABC + \bar{A}BC$
Solution: $A\bar{B}\bar{C} + BC$

1. Look at column 7. Draw a line through all rows whose terms are *not* contained in the expression to be simplified (rows 1, 2, 3, 6, and 7 in this example).
2. Starting with column 1, rule out all terms lined out in step 1. Note that $\bar{A}$ is lined out in row 4, and A is lined out in row 8. In column 2, B is lined out in rows 4 and 8, and $\bar{B}$ is lined out in row 5. Following the procedure in this example, all terms are crossed out in columns 1, 2, 3, 4, and 5.
3. In column 6, BC (rows 4 and 8) is not eliminated. Circle these for identification.

4. Starting at the left, go to the right and rule out all terms containing BC in all rows that contain a circled BC (in this example, rows 4 and 8, column 7).

5. All terms are now ruled out except BC in column 6, and $A\bar{B}\bar{C}$ in column 7. Only $A\bar{B}\bar{C}$ and BC remain. Therefore ABC + ABC + $\bar{A}$BC = $A\bar{B}\bar{C}$ + BC.

SUMMARY: Boolean relationships can be simplified by several means, i.e., algebraic, switch contacts, truth tables, Karnaugh maps, or charts.

DeMorgan's theorem (Rules 25 and 26 of Table 3-2) illustrate the *duality* of Boolean (hence logic) relationships. In words:

1. The truth value of statement A OR B is identical to the truth value of the *negation* of the statement A NOT AND B NOT. In symbols, $A+B = \overline{\bar{A}\bar{B}}$.

2. The truth value of statement A AND B is identical to the *negation* of the statement A NOT OR B NOT. In symbols, $AB = \overline{\bar{A}+\bar{B}}$

This can be proved by the truth table of Table 3-4. We will repeat here the rules for construction of such a truth table as applied specifically to Table 3-4.

1. List the variables and all possible combinations (row 1, columns 1 through 12 in this example).

2. Determine the values of elementary letters (rows 2 through 5, columns 1 through 4, in this example).

3. From step 2, determine the values of combinations (rows 2 through 5, columns 5 through 12, in this example).

Table 3-4. Proof of DeMorgan's Theorem by Truth Table

Row	Column											
	1	2	3	4	5	6	7	8	9	10	11	12
1	A	B	$\bar{A}$	$\bar{B}$	AB	$\overline{AB}$	$\bar{A}\bar{B}$	$\overline{\bar{A}\bar{B}}$	$A+B$	$\overline{A+B}$	$\bar{A}+\bar{B}$	$\overline{\bar{A}+\bar{B}}$
2	0	0	1	1	0	1	1	0	0	1	1	0
3	0	1	1	0	0	1	0	1	1	0	1	0
4	1	0	0	1	0	1	0	1	1	0	1	0
5	1	1	0	0	1	0	0	1	1	0	0	1

Note that

$$A+B = \overline{\overline{A}\overline{B}}$$
$$AB = \overline{\overline{A}+\overline{B}}$$
$$\overline{AB} = \overline{A}+\overline{B} \qquad \text{(Rule 25)}$$
$$\overline{AB} = \overline{A+B} \qquad \text{(Rule 26)}$$

Restate DeMorgan's theorem as follows: If every $+$ is replaced by a $\cdot$ and every $\cdot$ is replaced by a $+$, and each variable is replaced by its inverted value, then the resultant function equals this function inverted. (NOTE: In our treatment, the symbol $\cdot$ is implied, where $AB = (A \cdot B = $ "A AND B.")

Given: $f = \overline{AB}$. Derive DeMorgan's theorem, Rule 25.

Step 1. Replace implied sign $\cdot$ with $+$ as: $\overline{A+B}$.

Step 2. Invert each variable: $\overline{A}+\overline{B}$.

Step 3. Then function $f = \overline{A}+\overline{B} = \overline{AB}$. This is derived Rule 25, illustrated by Fig. 3-1, row 8.

Given: $f = \overline{A+B}$. Derive DeMorgan's theorem, Rule 26.

Step 1. Replace $+$ with $\cdot$ (implied): $\overline{AB}$.

Step 2. Invert each variable: $\overline{A}\overline{B}$.

Step 3. Then $f = \overline{A}\overline{B} = \overline{A+B}$. This is derived Rule 26, illustrated by Fig. 3-1, row 4.

3-3. LOGIC DESIGN

From Table 1-8 (Chapter 1) and associated text, the Boolean relationship for binary sum is:

$$\text{Sum} = \overline{A}\overline{B}C + \overline{A}B\overline{C} + A\overline{B}\overline{C} + ABC$$

and also that Boolean for the carry output is:

$$C_n = \overline{A}BC + A\overline{B}C + AB\overline{C} + ABC.$$

If we were to construct a full binary adder from the Boolean relationships above, we would have the circuitry of Fig. 3-4. Note that three inverter gates, eight AND gates, and two OR gates are required, for a total of thirteen symbols. This is *not* a valid design as proved by the following text.

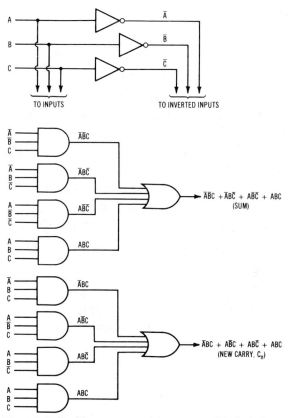

Fig. 3-4. Full adder constructed from unsimplified Boolean
relationships of Table 1-8.

Let us simplify the above Boolean notations algebraically, and
prove this simplification by a truth table. First, we will use the
algebraic method:

$$\text{Sum} = \bar{A}\bar{B}C + \bar{A}B\bar{C} + A\bar{B}\bar{C} + ABC$$
$$= C(\bar{A}\bar{B}) + \bar{C}(\bar{A}B) + \bar{C}(A\bar{B}) + C(AB) \qquad \text{Rules 18, 20}$$
$$= C(AB+\bar{A}\bar{B}) + \bar{C}(\bar{A}B+A\bar{B}) \qquad \text{Rule 21}$$

$$C_n = \bar{A}BC + A\bar{B}C + AB\bar{C} + ABC$$
$$= C(\bar{A}B) + C(A\bar{B}) + \bar{C}(AB) + C(AB) \qquad \text{Rules 18, 20}$$
$$= C(\bar{A}B+A\bar{B}) + AB \qquad \text{Rule 21}$$
$$= AB+C(A\bar{B}+\bar{A}B) \qquad \text{Rule 17}$$

So now we can put down the simplified Boolean relationships as follows:

$$\text{Sum} = C(AB + \bar{A}\bar{B}) + \bar{C}(\bar{A}B + A\bar{B})$$

$$C_n = AB + C(A\bar{B} + \bar{A}B)$$

At this time, review the instructions for compiling the truth tables of Tables 3-3 and 3-4 in Section 3-2. Use these procedures in constructing the truth table for binary sum and carry as in Table 3-5. Note carefully how this fits the binary addition rules for sum and carry.

Sum = 1 when a single 1 occurs in A, B, and C. Carry = 0.
Sum = 0 when two 1s occur in A, B, and C. Carry = 1.
Sum = 1 when three 1s occur in A, B, and C. Carry = 1.

Decision-making logic is based upon exactly the same principles. Assume the sum represents a group of three objects (A,B,C), and at least two-thirds of each group must meet a specification represented by logical 1. The carry output would then represent the decision: if only 1 of the group is logical 1, the logical 0 output rejects the group. If two or more of the group are logical 1, the logical 1 output accepts that group. The entire logical operation here would involve only the carry circuitry.

Full Adder From Simplified Relationship

Now let's construct a full adder from the above simplified relationships for Boolean sum and carry output, as shown by Fig. 3-5. Here we have:

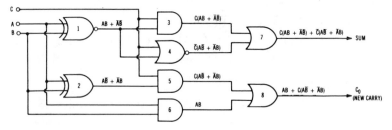

Fig. 3-5. Full adder constructed from simplified Boolean relationships.

Gate 1: exclusive NOR
Gate 2: exclusive OR
Gates 3, 5, 6: AND
Gate 4: NOR
Gates 7, 8: OR

This requires only 8 symbols in contrast to 13 symbols for Fig. 3-4. This is only one of many possible combinations.

Dual Solutions

There may be more than one correct solution to a simplification of relationships. For example, take the carry output (C_n) of Fig. 3-4:

$$C_n = \bar{A}BC + A\bar{B}C + AB\bar{C} + ABC$$
$$= \bar{A}BC + A\bar{B}C + AB(\bar{C} + C)$$
$$= \bar{A}BC + A\bar{B}C + AB(1) = \bar{A}BC + A\bar{B}C + AB$$
$$= \bar{A}BC + A(\bar{B}C + B) = \bar{A}BC + A(C + B) \quad \text{Rule 23}$$
$$= \bar{A}BC + AC + AB = B(\bar{A}C + A) + AC$$
$$= B(C + A) + AC = BC + AB + AC$$
$$= AB + AC + BC \quad\quad\quad\quad\quad \text{Rule 23}$$

For the carry logic, this would require the circuit of Fig. 3-6, replacing gates 2, 5, 6, and 8 of Fig. 3-5. Note carefully that this does not necessarily simplify Fig. 3-5; it only illustrates that more than one simplification procedure exists for a given set of terms, hence circuits can vary for identical functions.

You can always double-check any algebraic simplification by

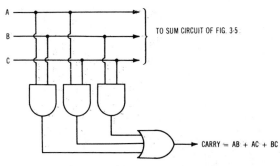

Fig. 3-6. Alternate carry circuit.

Table 3-5. Truth Table for Sum and Carry

A	B	C	$\bar{A}$	$\bar{B}$	$\bar{C}$	AB	$A\bar{B}$	$\bar{A}\bar{B}$	$\bar{A}B$	Sum		ORed Sum	Carry
										$C(AB + \bar{A}\bar{B})$	$\bar{C}(A\bar{B} + \bar{A}B)$	Sum	$AB + C(A\bar{B} + \bar{A}B)$
0	0	0	1	1	1	0	0	1	0	0	0	0	0
0	0	1	1	1	0	0	0	1	0	1	0	1	0
0	1	0	1	0	1	0	0	0	1	0	1	1	0
0	1	1	1	0	0	0	0	0	1	0	0	0	1
1	0	0	0	1	1	0	1	0	0	0	1	1	0
1	0	1	0	1	0	0	1	0	0	0	0	0	1
1	1	0	0	0	1	1	0	0	0	0	0	0	1
1	1	1	0	0	0	1	0	0	0	1	0	1	1

Table 3-6. Truth Table Proving Validity of AB + AC + BC for Carry

Binary				Binary Sum	Binary Carry
A	B	C	Boolean	(Table 3-5)	(AB+AC+BC)
0	0	0	$\bar{A}\bar{B}\bar{C}$	0	0
0	0	1	$\bar{A}\bar{B}C$	1	0
0	1	0	$\bar{A}B\bar{C}$	1	0
0	1	1	$\bar{A}BC$	0	1
1	0	0	$A\bar{B}\bar{C}$	1	0
1	0	1	$A\bar{B}C$	0	1
1	1	0	$AB\bar{C}$	0	1
1	1	1	ABC	1	1

means of a truth table. Table 3-6 proves the validity of AB+AC+ BC for carry. Therefore Figs. 3-5 and 3-6 are equally valid. The unsimplified version (Fig. 3-4) *would not be valid.*

Constructing a Full Binary Subtractor

Review Table 1-6 (Chapter 1) and draw up a new truth table for full binary subtraction that will include the Boolean expressions for f=1, as in Table 3-7. Note that

$$D = \bar{A}\bar{B}C + \bar{A}B\bar{C} + A\bar{B}\bar{C} + ABC$$
$$B_n = \bar{A}\bar{B}C + \bar{A}B\bar{C} + \bar{A}BC + ABC$$

where

B_n = new borrow,
C = previous borrow,
D = difference.

Note carefully that the truth table for the *difference* (D) column is identical to the *sum* column of Table 1-8 (Chapter 1).

We know now that before designing the full subtractor logic we will need to simplify the above relationships. So:

$$D = \bar{A}\bar{B}C + \bar{A}B\bar{C} + A\bar{B}\bar{C} + ABC$$
$$= C(AB+\bar{A}\bar{B}) + \bar{C}(A\bar{B}+\bar{A}B) \qquad \text{Rules 20, 21}$$
$$B_n = \bar{A}\bar{B}C + \bar{A}B\bar{C} + \bar{A}BC + ABC$$
$$= \bar{A}B + C(AB+\bar{A}\vec{B}) \qquad \text{Rules 20, 21}$$

Table 3-7. Full Subtractor Truth Table

A	B	C	D	B_n	f = 1 D	f = 1 B_n
0	0	0	0	0	0	0
0	0	1	1	1	$\bar{A}\bar{B}C$	$\bar{A}\bar{B}C$
0	1	0	1	1	$\bar{A}B\bar{C}$	$\bar{A}B\bar{C}$
0	1	1	0	1	0	$\bar{A}BC$
1	0	0	1	0	$A\bar{B}\bar{C}$	0
1	0	1	0	0	0	0
1	1	0	0	0	0	0
1	1	1	1	1	ABC	ABC

The difference (D) output is identical to that of the sum of Table 3-5 and Fig. 3-5. This function is based on the fact that A−B (A minus B) is identical with A+(−B), which reads "A plus minus B."

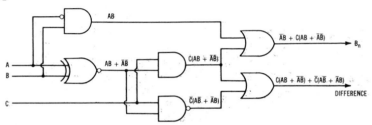

Fig. 3-7. Full binary subtractor.

Fig. 3-7 is one possible logic network to function as full binary subtractor. The functions A + B and A − B are identical for two bits. The resultant functional difference between add and subtract is the action of the previous carry (for add) and the previous borrow for subtract. Carefully compare Fig. 3-7 with Fig. 3-5 and note how this relationship is used.

3-4. THE IMPORTANCE OF SIMPLIFICATION

Suppose you have evolved the following Boolean expression for a given problem: $AC+ABC+A\bar{C}$.

Simplify this relationship, first by algebraic means and then by a truth table. To solve algebraically:

$$f = AC + ABC + A\bar{C}$$

$= A(C+\bar{C}) + ABC$		Rules 17, 21
$= A(1) + ABC$		Rule 12
$= A+ABC$		Rule 16
$= A$		Rule 22

This shows that the circuitry $AC+ABC+A\bar{C}$ can be replaced by a straight wire from A.

Now draw up the truth table for the same relationship following the rules as reviewed by Table 3-8. Note that the first and last columns are identical, proving that $AC+ABC+A\bar{C} = A$.

Table 3-8. Truth Table Method of Simplifying AC + ABC + AC̄

A	B	C	C̄	AC	AC̄	ABC	AC + ABC + AC̄
0	0	0	1	0	0	0	0
0	0	1	0	0	0	0	0
0	1	0	1	0	0	0	0
0	1	1	0	0	0	0	0
1	0	0	1	0	1	0	1
1	0	1	0	1	0	0	1
1	1	0	1	0	1	0	1
1	1	1	0	1	0	1	1

Not so apparent from the Boolean-binary section of Table 3-2 are certain inverted values of the variables A, B, etc., but these should be apparent from inspection. For example:

Rule 13: $\qquad\qquad 0+A = A$

The same relationship holds true for all variables. The variable A can also be an inverted variable, such as $\bar{B}$. Then the result is $0 + \bar{B} = \bar{B}$. Here $\bar{B}$ has replaced A in the theorem $0+A = A$. Remember that $\bar{B}$ can be either a 0 or a 1; therefore, if $\bar{B} = 0$, we have $0+0 = 0$. If $\bar{B} = 1$, we have $0+1 = 1$.

Similarly, Rule 14, $(0)(A) = 0$, could be $(0)(\bar{B}) = 0$. In the AND operation, the result, or function, is always 0 if any variable is a 0, e.g., $(0)(1) = 0$.

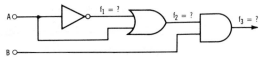

Fig. 3-8. Circuit for Exercise 3-5.

EXERCISES

3-1. Write the Boolean relationship for the following statement: "The mission will proceed if Jack and Jill are both present, and it is not snowing."

3-2. Draw up the truth table for the answer to Exercise 3-1.

3-3. Prove algebraically that $A\bar{B}0 = 0$ (read as A AND B NOT AND zero equals zero).

3-4. Prove algebraically that $A+A+\bar{A}+1 = 1$.

3-5. In Fig. 3-8, fill in the values of f_1 and f_2 in simplest terms. Then what is the functional output f_3?

3-6. In Fig. 3-9, fill in the values of f_1 through f_5, simplifying as you go along. Then what is the value of the functional output f_6?

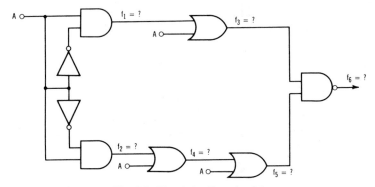

Fig. 3-9. Circuit for Exercise 3-6.

3-7. Draw a simplified schematic that would give the same function as the answer to Exercise 3-6.

3-8. Prove Rule 23 (Table 3-2) by reasoning aloud.

3-9. Apply DeMorgan's theorem, Rule 26, to the expression $\overline{\bar{A}+B+C}$.

3-10. Simplify algebraically the expression $f = AC+AD+BC+BD$.

CHAPTER 4

Boolean Algebra in Pictorial Form

Pictorial representations of Boolean relationships, known as Venn diagrams, are useful not only as one form of simplification, but also as direct practice in logical thinking. The Venn diagram illustrates algebra applied to classes.

4-1. CLASSES IN PICTORIAL FORM

The rectangle of Fig. 4-1A represents a universe of cars. The universe of cars is the class of all cars. A *universe* is the totality of objects under consideration. For example, the universe of people is the class of all people. The universe of colors is the class containing all colors.

(A) Rectangle represents a universe of cars.

(B) Shaded area A represents red cars.

Fig. 4-1. Classification of cars as example.

If the area within the rectangle of Fig. 4-1A represents the universe class, then it is designated by 1. The area outside this rectangle is not a member of the universe class and is called the *null class,* designated by 0. This division illustrates the axioms of previous study in this book that:

$$0+0 = 0 \qquad\qquad 0+1 = 1$$
$$0 \cdot 1 = 0 \qquad\qquad 1+1 = 1$$
$$0 \cdot 0 = 0 \qquad\qquad 1 \cdot 1 = 1$$

Let's explore the preceding two relationships that contain both a 0 and a 1; these are: $0+1 = 1$ and $0 \cdot 1 = 0$. The $0+1$ indicates a new class with members in the universe class *or* with members in the null class. But the null class has no members, so the new class can have only members from the universe class; therefore $0+1 = 1$. The class $0 \cdot 1 = 0$ shows the possibility of a class with members in both the universe *and* null class. Since these classes are mutually exclusive, this condition cannot be true. Thus $0 \cdot 1 = 0$. The class $1+1 = 1$ and the class $1 \cdot 1 = 1$ do exist as indicated by the logical result.

In Fig. 4-1B we have selected the class A in the universe class. If the shaded area A represents red cars, then the unshaded area within the rectangle represents the class of cars that are not red. The total rectangular area consists of the circle plus the area outside the circle but inside the rectangle. Therefore the area A plus the area $\bar{A}$ is the total area of the universe of cars. This proves that $A+\bar{A} = 1$.

The area outside the universe does not exist insofar as this universe is concerned. The null class always has the value of 0. Since the class of red cars (A) must be contained within the universe of cars, it cannot be placed outside the universe (in 0). Therefore, A and 0 have no common members. Thus $A \cdot 0 = 0$.

4-2. THE AND-OPERATION VENN DIAGRAM

In Fig. 4-2A consider the rectangle as some large collection under consideration, termed the "universal class" or simply the "universe." Subclasses A and B are to be examined. Since objects can be members of more than one class, A and B are variables. If A is the class of cars, and B is the class of red objects, then the new class (A AND B) is the class of red cars. The class A AND B (Fig. 4-2B) consists of those items that are members of both class A and class B. Since all points within circle A represent members of class A and all points within circle B represent members of class B, the shaded area AB in Fig. 4-2B is the intersection A AND B. (The shaded area is in both A AND B). In Venn diagrams the AND operation means *intersection:* the result of the operation is indicated by this intersection. Thus if class A = "cars" and class B = "objects that are red," class AB (A AND B) = "cars that are red."

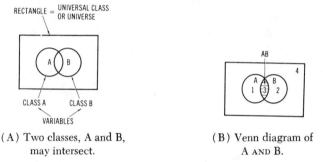

(A) Two classes, A and B, may intersect.

(B) Venn diagram of A AND B.

Fig. 4-2. Boolean operation AND is indicated by the intersection of two classes.

Table 4-1 shows the derivation of subclasses when intersecting classes A and B are considered. Since there are four subclasses of classes A and B (ways in which members of these classes can be represented), and since a complement exists for each of these

Table 4-1. The Eight Subclasses of A AND B

Row	Subclass	Area in Fig. 4-2B	Complement	OR-Function Equivalent or Complement
1	$\overline{A}\overline{B}$	4	$\overline{\overline{A}\overline{B}}$	$A + B$
2	$\overline{A}B$	2	$\overline{\overline{A}B}$	$A + \overline{B}$
3	$A\overline{B}$	1	$\overline{A\overline{B}}$	$\overline{A} + B$
4	AB	3	$\overline{AB}$	$\overline{A} + \overline{B}$

four subclasses, we have a total of eight subclasses for a two-variable function.

Proceeding to the right in row 1 of Table 4-1, we find the complement of $\bar{A}\bar{B}$, and you will recall that a complement of an AND function has an equivalent OR-function expression (DeMorgan's theorem, Chapter 3). Table 4-1 thus illustrates two expressions of each subclass in terms of A and B. In row 2 area 2 is derived, in row 3 area 1 is derived, and in row 4 area 3 is derived. The subclass, its complement (inversion), and the OR-function equivalent of its complement is indicated for each subclass.

Now assume we have an AND operation with one negation. For example, if A is classed as "cars" and B is the class of red objects, $A\bar{B}$ is the class of cars that are *not* red. This is represented by Fig. 4-3A. Note that only a portion of the A circle is shaded; the intersection of A and B is not shaded. The entire area within the rectangle but outside the B circle is $\bar{B}$.

(A) Shaded area is $f = A\bar{B}$.

(B) Shaded area is $f = \bar{A}B$.

Fig. 4-3. Boolean AND operation with one negation.

In Fig. 4-3B only a portion of the B circle is shaded; the *intersection* of $\bar{A}$ and B is thus $\bar{A}B$ (A NOT AND B). Thus we have the class of red objects that are *not* cars.

Consider for the moment a single NOT class within the universe of classes (Fig. 4-4A). In this case all the shaded area is NOT in A. Obviously, this could be done for any other single variable

(A) $f = \bar{A}$.

(B) $f = \bar{A}\bar{B}$.

(C) $f = \overline{AB}$.

Fig. 4-4. Complements, intersections of complements, and complement of an intersection.

such as B, C, X, Y, Z, etc. In Fig. 4-4B, the intersection A NOT AND B NOT is shaded to represent $\bar{A}\bar{B}$. In Fig. 4-4C, shading is in the area that is NOT in A AND B, to indicate $\overline{AB}$.

Venn diagrams can represent more than two variables, as indicated by Fig. 4-5. Fig. 4-5A shows the shaded area in the intersection of A AND B AND C; thus f = ABC. In Fig. 4-5B the shaded area is in A AND C, but NOT in A AND B or B AND C; thus f = $A\bar{B}C$ (read A AND B NOT AND C).

(A) f = ABC.　　　　　　(B) f = $A\bar{B}C$.

Fig. 4-5. Venn diagrams for three variables.

4-3. THE OR-OPERATION VENN DIAGRAM

Whereas the AND function for a Venn diagram is expressed in the *intersection* (superimposition) of distinct classes, the OR function is expressed in the *union* of distinct classes.

Fig. 4-6 shows the four initial subclasses and their complements, along with the appropriate Venn diagrams. Proceeding to the right in row 1, we find the inversion of $\bar{A}\bar{B}$ and we know that there exists an OR equivalent of the inverted AND function. This becomes very evident by Venn diagrams. Note that in the diagram for $\bar{A}\bar{B}$ the shading is in the area that is NOT in A AND NOT in B. Now if you simply invert the shading so that the circles representing A and B are shaded but the area outside the circles is not shaded, you have the Venn diagram for A+B (the OR equivalent of the inverted A NOT AND B NOT). Since the same shading is in both circles, we have a *union* representing the Boolean OR function.

Row 2 shows the initial subclass $\bar{A}B$, which is the same as Fig. 4-3B. If you reverse the area of shading, you have the Venn diagram for the OR equivalent of inverted $\bar{A}B$. Note that this is now a *union* representing A+$\bar{B}$. With this help, you should be able to analyze rows 3 and 4 in the same manner.

ROW	SUBCLASS	COMPLEMENT	NEW OR SUBCLASS	PICTORIAL	
				SUBCLASS (AND)	COMPLEMENTED SUBCLASS (OR)
1	$\bar{A}\bar{B}$	$\overline{\bar{A}\bar{B}}$	$A+B$	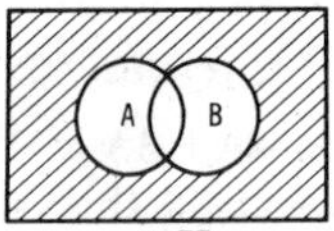 $f=\bar{A}\bar{B}$	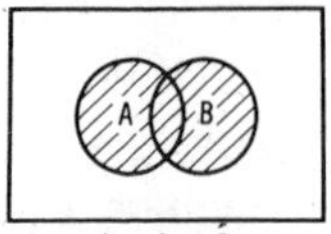 $f=A+B$
2	$\bar{A}B$	$\overline{\bar{A}B}$	$A+\bar{B}$	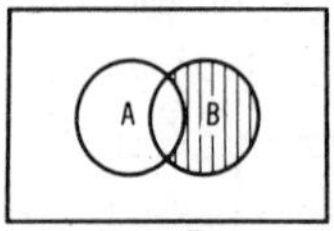 $f=\bar{A}B$	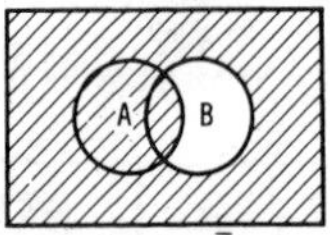 $f=A+\bar{B}$
3	$A\bar{B}$	$\overline{A\bar{B}}$	$\bar{A}+B$	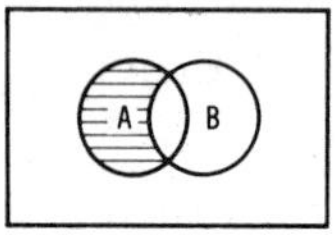 $f=A\bar{B}$	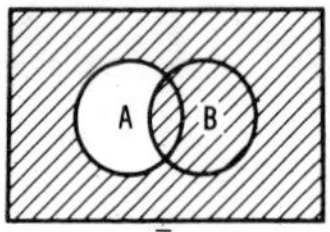 $f=\bar{A}+B$
4	AB	$\overline{AB}$	$\bar{A}+\bar{B}$	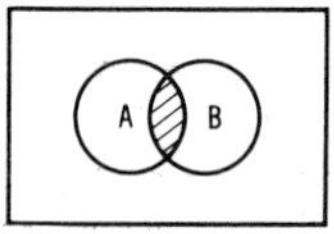 $f=AB$	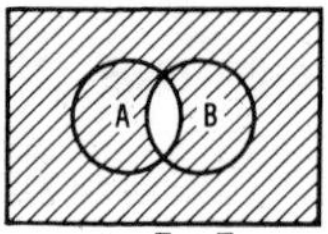 $f=\bar{A}+\bar{B}$

Fig. 4-6. The eight subclasses of classes A and B.

4-4. COMBINATIONAL VENN DIAGRAMS

If the superimposed (junction) area is shaded differently than the variables or the universe, we have a positive *junction*. (See Figs. 4-2B, 4-5A, 4-5B, and 4-6, row 4.) If the one or more classes in the junction are shaded the same, we have a positive *union* (Fig. 4-6, row 1, OR function). All other conditions are representative of at least one negative (inverted) variable in the expression, which has an equivalent as derived from DeMorgan's theorem. This can represent a combinational function, either an equivalent AND, and equivalent OR, or both.

Review Fig. 4-2B for the Venn diagram of AB. If you reverse (invert) this diagram, you have the OR function of Fig. 4-6, row 4. This says that $\overline{AB}$ (NOT A AND B, Fig. 4-4C) is equal to $\bar{A}+\bar{B}$ (A NOT OR B NOT) This Venn diagram can be read either way and proves DeMorgan's theorem. (Review Fig. 3-1, row 8.)

Review Fig. 4-4B and the AND function of Fig. 4-6, row 1. Note that the OR operation is just the inverse of the AND operation, and represents A+B, which is the equivalent of the *inverse* of $\overline{AB}$. (Reviewing Fig. 2-10B, you will note that if you invert $\overline{AB}$, you have the OR function A+B.) Thus if you again invert the OR function, you have the original AND function $\overline{AB}$, which is equal to the inverse of A+B, which is $\overline{A+B}$ (NOT A OR B). Thus the AND operation $\overline{AB}$ is equivalent to $\overline{A+B}$. (Review Fig. 3-1, row 4.) Again, this is a demonstration of DeMorgan's theorem.

As another example, assume you want to illustrate the function $A\overline{C} + \overline{A}BC$. Observing Fig. 4-7, note in step 1 that if the entire area of A was shaded, then the function(f) would simply equal A. However, the junction of A and C is not shaded, so $f = A\overline{C}$. In step 2, $\overline{A}BC$ is shown shaded in the region outside A but inside B AND C. In step 3, $A\overline{C} + \overline{A}BC$ is represented by the union of steps 1 and 2.

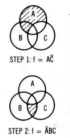

STEP 1: f = A$\overline{C}$

STEP 2: f = $\overline{A}$BC

STEP 3: f = A$\overline{C}$ + $\overline{A}$BC

Fig. 4-7. Development of $A\overline{C} + \overline{A}BC$.

Note that in the Venn diagrams we darken the areas corresponding to those class combinations for which the function has the value 1. You will see the similarity to a truth table if you use a shaded portion for the function 1. If there are n classes, then there will be 2^n subclasses, or class combinations.

Fig. 4-8 is detailed in a somewhat different way (as in Table 4-1) to clarify some of the functions. In (1), we review the function AB, which is illustrated by the shading in the *junction* of A AND B. In (2) of Fig. 4-8 we have the typical *union* of A and B, so f=A+B+AB (A AND/OR B) to illustrate the inclusive OR function. In (3) of Fig. 4-8 area A is drawn with horizontal lines and area B with vertical lines. This is a combination union and junc-

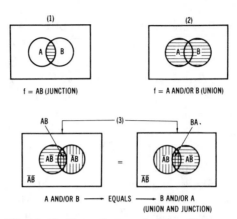

Fig. 4-8. Shading techniques for Venn diagrams.

tion (the junction is crosshatched) that illustrates the commutative law: A AND/OR B = B AND/OR A (Rules 17 and 18, Table 3-2).

4-5. SIMPLIFICATION BY VENN DIAGRAMS

Item (3) of Fig. 4-8 can be used to show the technique of simplification by using Venn diagrams. Recall that this may be done by algebraic means, by using a truth table, visually by Venn diagrams, and by using the Karnaugh map (covered in the next chapter).

The expression $A\bar{B} + AB$ can be shown to equal A. Algebraically: $A\bar{B} + AB = A(\bar{B}+B) = A(1) = A$

By using drawing (3) of Fig. 4-8, you can see that the area $A\bar{B}$ plus the area AB is the total area that constitutes the original region A. Thus this diagram shows the simplification of a Boolean statement: $A\bar{B} + AB = A$.

From this same drawing, you can also see the fact that Boolean Rule 22 (Table 3-2) is valid: $A+AB = A$. You will observe that area AB is a part of the circle representing A. Thus it is obvious that the area representing A plus the area AB is still the original area A.

Now observe Fig. 4-9 and note that three variables define $2^3 = 8$ classes as shown in Boolean form on the drawing. Class A is drawn with horizontal lines, class B with vertical lines and class

C with slanted lines. How many simplifications can you see on this drawing?

1. The original area A is the sum of the areas $A\bar{B}\bar{C}$ plus $AB\bar{C}$ plus ABC plus $A\bar{B}C$. Thus:
$$A\bar{B}\bar{C}+AB\bar{C}+ABC+A\bar{B}C=A$$

2. The original area B is the sum of the areas $\bar{A}B\bar{C}$, $\bar{A}BC$, ABC, and $AB\bar{C}$.

3. The original area C is the sum of areas $\bar{A}\bar{B}C$, $\bar{A}BC$, ABC, and $A\bar{B}C$.

Other simplifications exist but these are left to the imagination of the reader or instructor.

Table 4-2. Proof That $A\bar{B}\bar{C}$ + $AB\bar{C}$ + ABC + $A\bar{B}C$ = A

A	B	C	Ā	B̄	C̄	AB̄C̄	ABC̄	ABC	AB̄C	AB̄C̄ + ABC̄ + ABC + AB̄C
0	0	0	1	1	1	0	0	0	0	0
0	0	1	1	1	0	0	0	0	0	0
0	1	0	1	0	1	0	0	0	0	0
0	1	1	1	0	0	0	0	0	0	0
1	0	0	0	1	1	1	0	0	0	1
1	0	1	0	1	0	0	0	0	1	1
1	1	0	0	0	1	0	1	0	0	1
1	1	1	0	0	0	0	0	1	0	1

When you have simplified a Boolean statement by one method, always double-check by another method. For example, Statement 1 above says that:

$$A\bar{B}\bar{C} + AB\bar{C} + ABC + \bar{A}BC = A$$

This statement can be proved by another method such as using the truth table as in Table 4-2. Note that the first and last columns are identical. For the technique of composing such a truth table, review Section 3-2.

4-6. IMPORTANT REVIEW OF VENN DIAGRAMS

1. The primary terms are shaded in horizontal, vertical, or diagonal directions.

2. If the expressions are connected by the operator AND (multiplication sign · usually assumed), the crosshatched area is the area of interest.

3. If the expressions are connected by the operator OR (plus sign, +), the total shaded and crosshatched area is the area of interest.

The Venn diagram is handy to use for up to three variables. For more than three variables, use of the Karnaugh map is most convenient as discussed in the next chapter.

EXERCISES

4-1. Shade the proper areas for Boolean expression $A\bar{B} + B\bar{A}$ on a Venn diagram.

4-2. Shade the expression $AB + \bar{A}\bar{B}$ on a Venn diagram.

4-3. Draw $AB+C$ on a Venn diagram.

4-4. Draw $A+\bar{C}$ on a Venn diagram.

4-5. Draw AC on a Venn diagram.

4-6. Draw $AB\bar{C}$ on a Venn diagram.

4-7. Simplify the Boolean expression $\overline{AB} + B\bar{A}$ by using: (a) Boolean algebra and (b) a Venn diagram.

4-8. Simplify the Boolean expression $A\bar{B} + B\bar{A} + \overline{AB}$ by using: (a) Boolean algebra and (b) a Venn diagram.

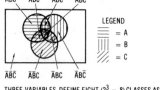

ABC̄   ĀBC̄   ABC̄   ABC

LEGEND
≡ = A
||| = B
/// = C

ĀB̄C̄   ĀB̄C   ĀBC   ĀBC̄

Fig. 4-9. Many equivalences can be found in this Venn diagram.

THREE VARIABLES DEFINE EIGHT (2^3 = 8) CLASSES AS FOLLOWS:

BOOLEAN	BINARY	BOOLEAN	BINARY
ĀB̄C̄	000	AB̄C̄	100
ĀB̄C	001	AB̄C	101
ĀBC̄	010	ABC̄	110
ĀBC	011	ABC	111

4-9. Prove by means of a truth table that the original area B in Fig. 4-9 is $\bar{A}B\bar{C} + \bar{A}BC + ABC + AB\bar{C}$.

4-10. Prove by means of a truth table that the original area C in Fig. 4-9 is $\bar{A}\bar{B}C + A\bar{B}C + ABC + \bar{A}BC$.

CHAPTER 5

The Karnaugh Map

A common and powerful tool used by the logic designer is called the Karnaugh map* or Veitch diagram.** Like the Venn diagram, the Karnaugh map (K-map for short) allows a visual representation of data, but in a more practical form to allow handling of up to four variables without difficulty. Up to six variables can be handled with more skill but become rather difficult. For more than four variables, other forms of simplification should be used. Two types of K-maps will be covered—the Veitch form and the binary form.

5-1. THE TWO-VARIABLE KARNAUGH MAP (VEITCH FORM)

The simplest map is the single-variable Boolean function of Fig. 5-1. Fig. 5-1A shows that the single variable can be either

---

* M. Karnaugh, "The Map Method for Synthesis of Combinational Logic Circuits," *Trans AIEE*, Vol. 72, Pt. I, pp. 593–598 (1953). This article expanded on the methods of the following reference and introduced a new form of chart.
** E. W. Veitch, "A Chart Method for Simplifying Truth Functions," *Proc. ACM*, Pittsburgh, PA, pp. 127–133 (May 1952).

A or $\overline{A}$ (1 or 0). Fig. 5-1B represents B and $\overline{B}$. To illustrate the similarity to a Venn diagram, note that (for example) the area of the rectangle representing the universe is the sum of A and $\overline{A}$, so that $A+\overline{A} = 1$. This could be shown by shading both halves of Fig. 5-1A. However, it is more effective if we consider the K-map as a pictorial presentation of the truth table such that functions can be pinpointed to a definite geographical area of a map. This will become more evident as we progress.

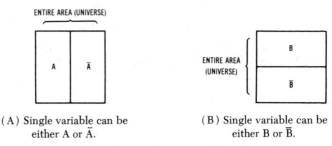

(A) Single variable can be either A or $\overline{A}$.

(B) Single variable can be either B or $\overline{B}$.

Fig. 5-1. A single variable can be either a 0 or a 1.

Development of a two-variable K-map of the Veitch form is shown in Fig. 5-2. For two variables, we must have $2^2 = 4$ primary values, hence four squares. Each of these areas is the result of superimposing the two single-variable diagrams of Fig. 5-1. Remember that on a Venn diagram the superimposition (junction) yields a Boolean AND statement. Fig. 5-2A therefore shows the obvious result of superimposing B and A of Fig. 5-1. This yields AB (A AND B). Similarly, Fig. 5-2B shows the area $\overline{A}$B. The other areas in Fig. 5-2 are produced in like manner, resulting in $A\overline{B}$ and $\overline{A}\overline{B}$.

Fig. 5-2E shows the final form of the two-variable K-map, with the numbers in the lower right-hand corner of each area representing the decimal equivalent of the binary value for easy identification of each area. If $A=1$, $\overline{A}=0$, and if $B=1$, $\overline{B}=0$. Then:

$\overline{A}\overline{B} = 00 = $ decimal 0
$\overline{A}B = 01 = $ decimal 1
$A\overline{B} = 10 = $ decimal 2
$AB = 11 = $ decimal 3

When variables A, B, etc., are designated as a 1 (no bar) or as a 0 (bar over letter), each of these is referred to as a *minterm*.

From Fig. 5-2E, note the obvious methods of simplification in Boolean statements. Look at the top row of Fig. 5-2E and note that the area consisting of AB plus ĀB is the area that was originally only B in Fig. 5-1. Remember that a *plus* statement is a Boolean or; therefore AB + ĀB = B (final simplification).

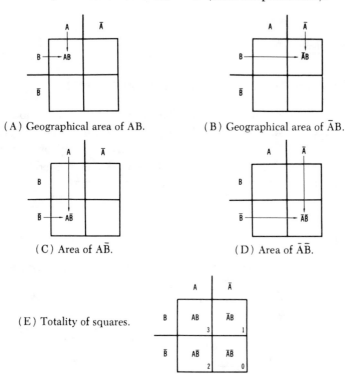

(A) Geographical area of AB.

(B) Geographical area of ĀB.

(C) Area of AB̄.

(D) Area of ĀB̄.

(E) Totality of squares.

Fig. 5-2. Development of a two-variable K-map.

When areas AB and ĀB (areas 1 and 3) were "picked up" to show that AB+ĀB = B, the process is termed a *coupling* of the two areas. Only horizontally or vertically adjacent areas can be coupled. In Fig. 5-2E, we can couple areas 3 and 1, 3 and 2, 2 and 0, or 0 and 1. It is *not correct* to couple areas that are diagonally positioned such as areas 3 and 0 or 1 and 2. Two areas can be coupled only if they give back the original area after coupling.

75

For example, if areas 2 and 3 are coupled, their sum will yield the original area A of Fig. 5-1. Thus:

$$A\bar{B} + AB = A$$

is a final simplification. However, if we sum the diagonal areas 1 and 2, the result does not yield any of the original areas. Therefore:

$$\bar{A}B + A\bar{B}$$

cannot be simplified further.

5-2. THE THREE-VARIABLE KARNAUGH MAP

When three variables exist, there are $2^3 = 8$ combination that can occur. Figs. 5-3A and 5-3B are repeats of Figs. 5-1A and 5-1B. In Fig. 5-3C, variable C is constructed vertically as was the diagram for variable A, except that the $\bar{C}$ area is divided into two

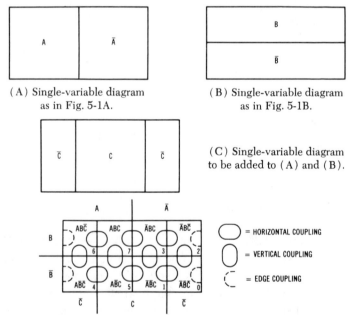

(A) Single-variable diagram as in Fig. 5-1A.

(B) Single-variable diagram as in Fig. 5-1B.

(C) Single-variable diagram to be added to (A) and (B).

(D) All possible values and couplings of three-variable K-map.

Fig. 5-3. Development of three-variable K-map.

parts as shown. Thus when variable C is superimposed on variables A and B, all eight possible combinations (8 squares) can be shown.

Fig. 5-3D shows all eight possible combinations of the three variables, as well as all possible couplings. As before, only horizontally adjacent or vertically adjacent areas can be coupled. But now we have an additional possibility. Note that since $\overline{C}$ is geographically placed at the extreme left and right edges of the map, it is possible to couple the edges of the diagram. You can visualize the map as being wrapped around a vertical cylinder such that the extreme edges of the map are adjacent to each other and therefore capable of being coupled. The technique of employing couples to simplify Boolean statements will be expanded as soon as we are more familiar with the development of K-maps for up to four variables.

5-3. THE FOUR-VARIABLE KARNAUGH MAP

See the "original diagram" part of Fig. 5-4 and note that the fourth variable (D) is placed such that the $\overline{D}$ area is divided into two parts: one on top and one on the bottom. When the variable D is superimposed on variables A, B, and C, all sixteen possible combinations ($2^4 = 16$) can be presented. Since D is divided into top and bottom, it is now possible to couple top and bottom areas of the map that become adjacent to each other if the map is considered wrapped around a horizontal cylinder.

First, let's see how easily Boolean statements with four variables can be simplified. Note, for example, that the eight areas on the left of the "original diagram" together equal the original area A (Section A of "extensions" in Fig. 5-4). Therefore, the sum of the areas is equal to A:

$$AB\overline{C}\overline{D} + ABC\overline{D} + ABCD + AB\overline{C}D + A\overline{B}\overline{C}D +$$
$$A\overline{B}CD + A\overline{B}C\overline{D} + A\overline{B}\overline{C}\overline{D} = A$$

Since Boolean statements normally involve no more than four variables, let's now take time to gain a practical insight as to how the K-map is used in logical circuit design.

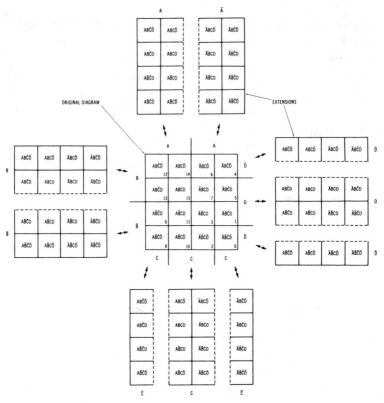

Fig. 5-4. Four-variable K-map, exploded view.

5-4. PRACTICAL USE OF THE K-MAP (VEITCH FORM)

First, we will review Fig. 5-2E. The area enclosed by the entire left hand side is A, while the right side describes Ā. The entire upper half is B, and the bottom half is B̄. With two variables A and B, one (either A or B) identifies two squares as a group; for example, A describes minterms 3 and 2 as a group. The two (A and B together) describe any one square; for example, AB̄ describes minterm 2 alone.

Now review Fig. 5-3D. There are three variables: A, B, and C. Any one variable describes four squares; for example, C describes minterms 1, 3, 5 and 7 as a group. Any two variables together describe two squares; for example, ĀC̄ describes minterms 0 and

2 as a group. Any three variables together describe a single square; thus $AB\overline{C}$ describes minterm 6 alone.

Next, review the original diagram of Fig. 5-4. We now have the four variables A, B, C, and D. Any one variable describes eight squares; for example, D describes squares 1, 3, 5, 7, 9, 11, 13, and 15 as a group. Any two variables together describe four squares; thus $\overline{A}C$ describes squares 2, 3, 6, and 7 as a group. Any three variables together describe two squares; for example, $A\overline{B}C$ describes squares 10 and 11 as a group. Any four variables together describe one square; $A\overline{B}CD$ describes square 11 alone.

Review all of the above until you are certain you understand the geographic relationship of the variables in the problem, since this leads to the following very important rule:

Rule: The number of different variables in the Boolean expression of the solution to a problem determines the minimum number of variables required to describe any pratircular group of squares in the K-map. The number of variables required to describe a single square is always exactly equal to the total number of different variables in the problem. Any lesser number of variables in an expression identifies only groups of squares, and the number of squares per group *increases* geometrically as the number of employed variables is *decreased* numerically.

An example of the above rule can be seen by observing the left vertical column of Fig. 5-4. This identifies the term $A\overline{C}$ (since $B+\overline{B} = 1$). Thus the simplification for $AB\overline{C}\overline{D} + AB\overline{C}D + A\overline{B}\overline{C}D + A\overline{B}\overline{C}\overline{D}$ is $A\overline{C}$. Note the extreme convenience of using the K-map for this simplification compared to the algebraic or any other method.

Now assume you want to simplify the Boolean expression:

$$f = AB\overline{C}\overline{D} + AB\overline{C}D + AB\overline{D} + \overline{A}BC\overline{D} + \overline{A}B$$

Use the following procedure:

1. Observe that the maximum number of variables is four; therefore you will need to construct a four-variable K-map containing 16 squares. But first, observe Table 5-1 and note how each term of the problem describes a specific number

of minterms or squares on the map. Observe how some squares are described more than once. In this case, squares 6 and 12 are identified twice. They need only be marked once with a 1 as described below.

Table 5-1. Minterms $AB\overline{C}\overline{D}$ + $AB\overline{C}D$ + $AB\overline{D}$ + $\overline{A}BC\overline{D}$ + $\overline{A}B$

Item	f (Boolean)	Number of Squares Identified by Term	Decimal Value of Minterms
1	$AB\overline{C}\overline{D}$	1	12
2	$AB\overline{C}D$	1	13
3	$AB\overline{D}$	2	12, 14
4	$\overline{A}BC\overline{D}$	1	6
5	$\overline{A}B$	4	4, 5, 6, 7

2. Construct your four-variable K-map as in Fig. 5-5A. Place a 1 in the geographic location for each of the terms in the problem as shown. The first term, $AB\overline{C}\overline{D}$, is represented by square 12. The second term, $AB\overline{C}D$, is represented by square 13. The third term, $AB\overline{D}$, is represented by squares 12 and 14, but 12 is already used so you simply insert a 1 in square 14. The fourth term, $\overline{A}BC\overline{D}$, is represented in square 6. The fifth term, $\overline{A}B$, is represented in four squares: 4, 5, 6, and 7. Square 6 is already used, so insert a 1 in squares 4, 5, and 7.

3. Study Fig. 5-5B and note that the first obvious coupling (grouping) of adjacencies is the horizontal coupling of the entire top row. Since $A+\overline{A} = 1$, the horizontal coupling of the top row is equivalent (equal) to $B\overline{D}$.

The second possible coupling is the four squares in the upper right-hand corner indicated in Fig. 5-5C. Since $D+\overline{D} = 1$ and $C+\overline{C} = 1$, the indicated grouping is equal to $\overline{A}B$.

The edge squares 4, 5, 12, and 13 are four adjacent squares when you consider the map wrapped around a vertical cylinder. This grouping may be seen to describe $B\overline{C}$ (see Fig. 5-5D).

Fig. 5-5E sums all the groupings in this example of the simplification procedure. Thus the simplified function is $f = B\overline{D} + \overline{A}B + B\overline{C}$.

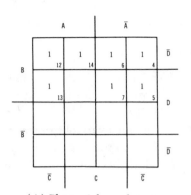

(A) Place a 1 for each term in the problem.

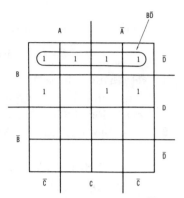

(B) First coupling is $B\overline{D}$.

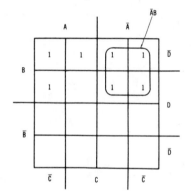

(C) Second coupling is $\overline{A}B$.

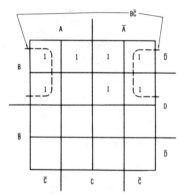

(D) Third (edge) coupling is $B\overline{C}$.

(E) Map showing all couplings.

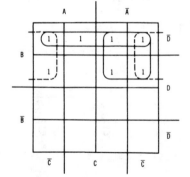

Fig. 5-5. Simplifying $AB\overline{C}\overline{D} + AB\overline{C}D + AB\overline{D} + \overline{A}B$ to $B\overline{D} + \overline{A}B + B\overline{C}$.

As another example, simplfy the following function:

$$f = A\bar{B} + B\bar{C} + \bar{B}C + \bar{A}B$$

1. Note that you have three variables (A, B, C) in this example. Construct a three-variable (eight-square) map for plotting. Review Fig. 5-3D.

2. See Table 5-2 and Fig. 5-6. For a three-variable function, any two variables together describe two squares. Note from Table 5-2 that, for example, $\bar{A}C$ describes squares 1 and 3 as a group.

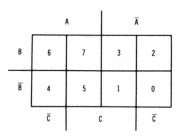

Fig. 5-6. Minterm values on K-map.

3. Taking the terms of the expression $f = A\bar{B}+B\bar{C}+\bar{B}C+\bar{A}B$, observe that:

$A\bar{B}$ describes squares 4 and 5.
$B\bar{C}$ describes squares 2 and 6.
$\bar{B}C$ describes squares 1 and 5.
$\bar{A}B$ describes squares 2 and 3.

Table 5-2. Values of Two- and Three-Variable Minterms

Minterms				Three-Variable Minterm Value*	
				Binary	Decimal
$\bar{A}\bar{B}\bar{C}$	$\bar{A}\bar{B}$	$\bar{A}\bar{C}$	$\bar{B}\bar{C}$	000	0
$\bar{A}\bar{B}C$	$\bar{A}\bar{B}$	$\bar{A}C$	$\bar{B}C$	001	1
$\bar{A}B\bar{C}$	$\bar{A}B$	$\bar{A}\bar{C}$	$B\bar{C}$	010	2
$\bar{A}BC$	$\bar{A}B$	$\bar{A}C$	BC	011	3
$A\bar{B}\bar{C}$	$A\bar{B}$	$A\bar{C}$	$\bar{B}\bar{C}$	100	4
$A\bar{B}C$	$A\bar{B}$	AC	$\bar{B}C$	101	5
$AB\bar{C}$	AB	$A\bar{C}$	$B\bar{C}$	110	6
ABC	AB	AC	BC	111	7

* See Fig. 5-6 for placement of minterm values.

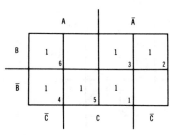

(A) Placing 1s in appropriate squares.

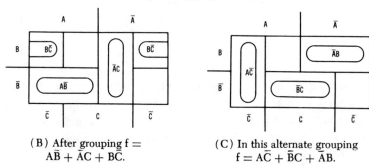

(B) After grouping f = $A\bar{B} + \bar{A}C + B\bar{C}$.

(C) In this alternate grouping f = $A\bar{C} + \bar{B}C + \bar{A}B$.

Fig. 5-7. Valid simplifications of f = $A\bar{B} + B\bar{C} + \bar{B}C + \bar{A}B$.

4. Place a 1 on the appropriate squares of the three-variable K-map as in Fig. 5-7A.

5. See Fig. 5-7B for one possible coupling to show simplified groupings of the original problem. Note that edge squares 2 and 6 are coupled when the map is considered wrapped around a vertical cylinder. This is equivalent to $B\bar{C}$. Vertical squares 1 and 3 are coupled to represent $\bar{A}C$. This leaves only horizontal squares 4 and 5 to be coupled, which become equivalent to $A\bar{B}$. Thus, in this grouping, f = $A\bar{B} + \bar{A}C + B\bar{C}$.

6. It is obvious from observation that more than one method of grouping is possible. This means that it is possible that more than one solution for a problem will give the same minimum number of variables. An alternate grouping is shown in Fig. 5-7C. In this case, f = $A\bar{C} + \bar{B}C + \bar{A}B$. Either solution is valid for design purposes.

As still another example, simplify f = $\bar{A}BC\bar{D} + \bar{A}\bar{B}C\bar{D}$. See the four-variable K-map of Fig. 5-8 for solution. Remember that

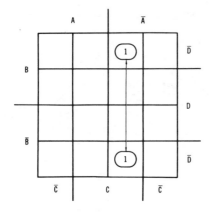

Fig. 5-8. Simplifying f =
$\bar{A}BC\bar{D} + \bar{A}BC\bar{D}$
to f = $\bar{A}C\bar{D}$.

since $\bar{D}$ is on both top and bottom, you can consider the map wrapped around a *horizontal* cylinder so that adjacencies can occur at top and bottom. Since $B+\bar{B} = 1$, the simplified function is independent of variable B and becomes f = $\bar{A}C\bar{D}$ as is obvious on the K-map.

5-5. ALTERNATE FORM OF K-MAP

The form of Karnaugh map described in previous sections is known as the Venn form or Veitch diagram where the areas originally were shaded or crosshatched as in a Venn diagram. For our purposes it is more practical to use Boolean notation as was carried out thus far. Another form of K-map attributed to Karnaugh alone uses only 0s and 1s rather than Boolean letter notation. This form is very popular today in design practice, and

A	B
0	0
0	1
1	0
1	1

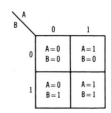

(A) Truth table form. (B) Two-variable K-map.

Fig. 5-9. Alternate form of two-variable K-map using binary 0s and 1s.

you should become familiar with both forms so that you can use the form easiest for you or for a particular problem.

The truth table form for two variables is reviewed in Fig. 5-9A. Note that this is transferred to Fig. 5-9B such that the 0 or 1 level of each variable is related to a given square on the map. In this example, A is represented horizontally and B is represented vertically. Note how the top left square therefore represents A=0, B=0. The top right square represents A=1, B=0. The bottom left square represents A=0, B=1, and the bottom right square represents A=1, B=1.

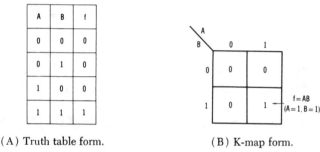

(A) Truth table form.　　　　　(B) K-map form.

Fig. 5-10. Truth table and K-map form of f = A and B.

In this section we will study how to plot the values on this form of K-map. The next section will cover its use in simplification procedure.

The truth table of Fig. 5-10A shows an output (binary 1) only when A and B are both high. We know this is an and function, and Fig. 5-10B shows how this is indicated on the K-map. The 1 occurs *only* in the square representing A and B.

The truth table of Fig. 5-11A indicates an or function since a high occurs when either or both of the variables are high. Fig. 5-11B indicates the same function on the K-map. The K-map is really a condensed form of truth table.

From the truth table of Fig. 5-12A, we note that a 1 occurs only when the variables are of opposite levels. This indicates the exclusive or function, and Fig. 5-12B is the K-map presentation.

The three-variable K-map is shown by Fig. 5-13. There are eight squares in the K-map, just as there are eight rows in a three-variable truth table. In Fig. 5-13A, AB is represented horizontally while C is represented vertically. In Fig. 5-13B, variable A is represented horizontally while BC is represented vertically. These two forms give identical results, and you will find both forms in use. We will use them interchangeably to provide practice in K-map reading.

A	B	f
0	0	0
0	1	1
1	0	1
1	1	1

(A) Truth table for f = A + B.

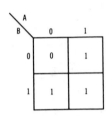

(B) K-map for f = A + B.

Fig. 5-11. Truth table and K-map of the OR function.

You should note an important point at this time. The sequence of binary values across the top row (horizontal) in Fig. 5-13A, or vertically in Fig. 5-13B, is not the same as in the conventional truth table where we simply add 1 to the preceding number. In switching from 1 to 2 (binary 01 to 10), both variables change simultaneously. The normal sequence is 00, 01, 10, 11. In Fig. 5-13, however, the sequence is 00, 01, 11, 10. Note that only one variable changes on each switching operation. This is brought to your attention at this time, and will be put to use later in your studies.

A	B	f
0	0	0
0	1	1
1	0	1
1	1	0

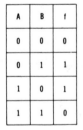

(A) Truth table for f = A$\overline{\text{B}}$ + $\overline{\text{A}}$B.

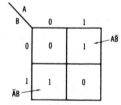

(B) K-map showing f = A$\overline{\text{B}}$ + $\overline{\text{A}}$B.

Fig. 5-12. The exclusive OR function.

Locate the square where A=1, B=0, and C=1. In Fig. 5-13A, this must be in the column where A=1 and B=0 (A$\bar{\text{B}}$), and in the row where C=1; this is in the lower right-hand square (A$\bar{\text{B}}$C). In Fig. 5-13B, it must be in the column where A=1, and in the row where B=0 and C=1. This occurs in the square designated A$\bar{\text{B}}$C.

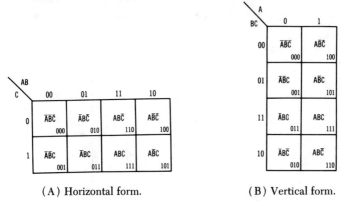

(A) Horizontal form.   (B) Vertical form.

Fig. 5-13. Two equivalent forms of three-variable Karnaugh maps.

The truth table of Fig. 5-14A indicates a "1" output under four specific levels of the three variables A, B, and C. This occurs for 001, 010, 011, and 101. Transfer the 1 outputs to the K-map of Fig. 5-14B. You know that there will be four 1s on the map. For example, in row 2 of the truth table, there is a 1 when A and B are 0 and C = 1. This is placed in the lower left-hand corner of Fig. 5-14B. In Fig. 5-14C, the 1 again occurs when A = 0, and when B = 0 and C = 1. This occurs in the second square from the top of the left-hand column. Place a 0 in all squares that do not have a binary 1.

A four variable K-map (Fig. 5-15) has sixteen squares arrange in four rows and four columns. Observe that the order of the rows is the same as the order of the columns: 00, 01, 11, 10.

One example of locating a given square is shown on Fig. 5-15. As another example, locate the square where A=0, B=0, C=1, and D=0. This must be in the column where A=0 and B=0. That column is 00. It must also be in the row where C=1 and D=0. That row is 10. The intersection of the two is the desired square

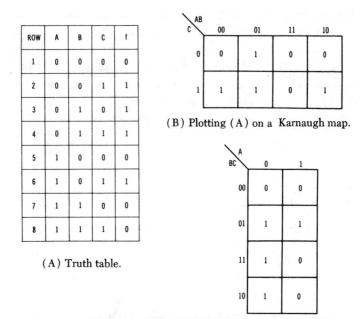

ROW	A	B	C	f
1	0	0	0	0
2	0	0	1	1
3	0	1	0	1
4	0	1	1	1
5	1	0	0	0
6	1	0	1	1
7	1	1	0	0
8	1	1	1	0

(A) Truth table.

(B) Plotting (A) on a Karnaugh map.

(C) Plotting (A) on other Karnaugh map.

Fig. 5-14. Plotting a truth table on two forms of Karnaugh maps.

$\overline{A}\overline{B}C\overline{D}$, or lower left-hand corner of Fig. 5-15. First, find the correct column, and then the correct row.

The truth table of Fig. 5-16A indicates a high level (binary 1) will occur at all times except when all variables are alike, either 0s or 1s. This information is transferred to the four-variable K-map

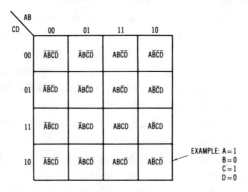

Fig. 5-15. Four-variable Karnaugh map with Boolean notation.

A	B	C	D	f
0	0	0	0	0
0	0	0	1	1
0	0	1	0	1
0	0	1	1	1
0	1	0	0	1
0	1	0	1	1
0	1	1	0	1
0	1	1	1	1
1	0	0	0	1
1	0	0	1	1
1	0	1	0	1
1	0	1	1	1
1	1	0	0	1
1	1	0	1	1
1	1	1	0	1
1	1	1	1	0

(A) Truth table.

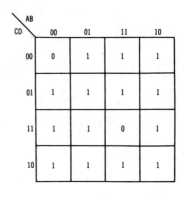

(B) Karnaugh map corresponding to (A).

Fig. 5-16. Plotting of truth table on Karnaugh map.

as in Fig. 5-16B. Note that a 0 occurs when all variables are 0 (upper left-hand corner) and when all variables are high (binary 1111). Place a 0 in these two squares and a 1 in all other squares.

Usually, since only the 1s are significant, the 0s are omitted. For example, plot f = $A\bar{B}CD$ + $\bar{A}BC$ (see Fig. 5-17). First, locate all squares that correspond to $\bar{A}BC$. Observe that $\bar{A}B$ is column 01. Go down this column until you intersect rows that correspond to C=1. These are 11 and 10; note that there are two squares involved in this example. Put a 1 in each square. Then locate the column corresponding to $A\bar{B}$. This is headed 10. Go down this column until you intersect the row CD (binary 11).

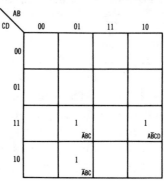

Fig. 5-17. Plot of $A\bar{B}CD$ + $\bar{A}BC$ on Karnaugh map, with 0s omitted.

Place a 1 in this square (A$\bar{B}$CD). This completes the K-map for A$\bar{B}$CD + $\bar{A}$BC.

5-6. SIMPLIFICATION PROCEDURES

The basic properties of any Karnaugh map are as follows:

1. Each square on the map relates to one specific value for each variable.
2. For adjacent squares, only one variable is different.
3. Adjacent squares have one common side. Diagonal squares are *not* adjacencies.

Observe Fig. 5-18 for a review of what constitutes "adjacencies" on the K-map. Remember that squares on the edges of each row (and each column) are adjacent. Prove this to yourself by

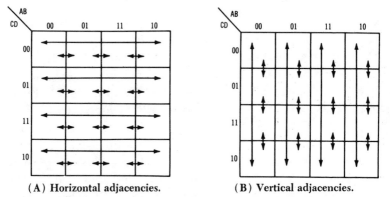

(A) Horizontal adjacencies.　　(B) Vertical adjacencies.

Fig. 5-18. The squares on the edge of each row (horizontal), and the squares on the edge of each column (vertical), are adjacent.

taking, for example, the edges of the top row where A=0, B=0, and D=0 (on the left) and A=1, B=0, C=0, and D=0 on the right edge. Only one variable (A in this example) is different (Rule 2 above). The same holds true for each adjacency within the outer boundary of the K-map.

Basic Rule: Two adjacent squares can be read as one term by including only those variables whose value is the same for both squares. The single variable that is different (opposite in value)

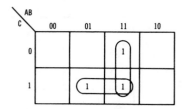

Fig. 5-19. Simplifying $\bar{A}BC +$ $AB\bar{C} + ABC$ to $AB + BC$.

is eliminated. For example, if you have $AB\bar{C}D$ and $\bar{A}B\bar{C}D$, the final term is $B\bar{C}D$ since $A + \bar{A} = 1$.

Review Fig. 5-17 and observe the position of the 1s for $\bar{A}BC$. Note that if all four variables are involved, we have $\bar{A}BCD$ and $\bar{A}BC\bar{D}$. Since the D term is of opposite value in the two squares, it is eliminated. This leaves only the variables that have the same value in each square. A single two-square coupling therefore eliminates one variable.

Example 1: Simplify $\bar{A}BC + AB\bar{C} + ABC$.

Solution: There are three variables, so plot the 1s on a three-variable K-map as in Fig. 5-19. In this example we can couple once vertically and once horizontally. (Note that the loops overlap.) The vertical coupling gives $AB\bar{C}$ with ABC. Since C and $\bar{C}$ occur, they are eliminated, giving AB. The horizontal coupling gives $\bar{A}BC$ with ABC. Since the terms A cancel, we have BC. Therefore the final simplification is $f = AB + BC$.

Example 2: Simplify $\bar{A}\bar{B}\bar{C} + \bar{A}B\bar{C} + A\bar{B}\bar{C} + A\bar{B}C$.

Solution: Plot the 1s of the terms in the problem as in Fig. 5-20. Note that the horizontal grouping is $\bar{A}\bar{B}\bar{C}$ with $\bar{A}B\bar{C}$ are equal to $\bar{A}\bar{C}$ (the B terms are different). The vertical grouping is $A\bar{B}\bar{C}$ with $A\bar{B}C$ are equal to $A\bar{B}$ (the C terms are different). Therefore $\bar{A}\bar{B}\bar{C} + \bar{A}B\bar{C} + A\bar{B}\bar{C} + A\bar{B}C = \bar{A}\bar{C} + A\bar{B}$ (final simplification).

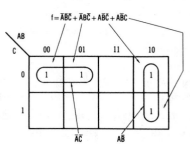

Fig. 5-20. Simplifying $f =$ $\bar{A}\bar{B}\bar{C} + \bar{A}B\bar{C} + A\bar{B}\bar{C} + A\bar{B}C$ to $\bar{A}\bar{C} + A\bar{B}$.

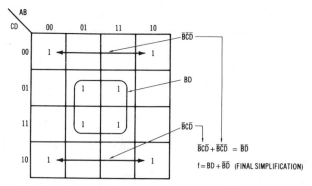

Fig. 5-21. Simplification of four center squares and four edge squares.

Example 3: Assume you have 1s positioned on a K-map as in Fig. 5-21. Note that a group of four 1s that is adjacent can be combined. Put down the values of these four 1s for each individual square: $\bar{A}B\bar{C}D$, $AB\bar{C}D$, $\bar{A}BCD$, $ABCD$. In this, note that the first two terms reduce to $B\bar{C}D$, and the second two terms reduce to BCD. These reduced terms are coupled and reduce to BD. So the four adjacent squares reduce to BD. A block of four squares eliminates two variables.

Now couple the two outer edges of the top row. Here we have $\bar{A}\bar{B}\bar{C}\bar{D}$ with $A\bar{B}\bar{C}\bar{D}$, or $\bar{B}\bar{C}\bar{D}$. (The A terms cancel.) Then couple the two outer edges of the bottom row. This is $\bar{A}\bar{B}C\bar{D}$ with $A\bar{B}C\bar{D}$ gives $\bar{B}C\bar{D}$. (The A terms cancel.) Combining these two simplifications, $\bar{B}\bar{C}\bar{D}$ with $\bar{B}C\bar{D}$, gives $\bar{B}\bar{D}$ (the C terms cancel). Thus the final simplification for this example is $f = BD + \bar{B}\bar{D}$.

You will recall that on the four-variable K-map, top and bottom edges (vertical) may also be coupled. This is an option you have in Fig. 5-21. In the left column you have $\bar{A}\bar{B}\bar{C}\bar{D}$ with $\bar{A}\bar{B}C\bar{D}$, $\bar{A}\bar{B}\bar{D}$. In the right column is $A\bar{B}\bar{C}\bar{D}$ with $A\bar{B}C\bar{D}$, or $A\bar{B}\bar{D}$. Combining the two reduced terms, $\bar{A}\bar{B}\bar{D}$ with $A\bar{B}\bar{D}$, gives $\bar{B}\bar{D}$. This is the same result as was obtained with horizontal edge coupling above.

A rule for any K-map is to go through the squares one at a time to see if they can be covered by one or more loops. Use the largest loop possible, which means the loop with the largest number of adjacent 1s. Only 2^n (n = 1, 2, 3, . . .) variables can be contained in a loop. If a 1 can obviously be covered in only one way, use that loop and go on. If it can be covered by more than one loop, study the overall map to ascertain the best grouping possible that will give the minimum number of loops (maximum number of 1s in each loop).

For example, how would you loop the map of Fig. 5-22A? One way (a wrong way) is shown in Fig. 5-22B. This shows four loops and therefore four groups of ORED terms each having three variables. The correct looping for this example is in Fig. 5-22C. Here there are only three loops and therefore three groups

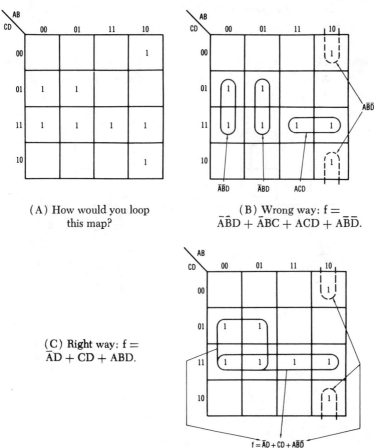

(A) How would you loop this map?

(B) Wrong way: f = $\overline{A}\overline{B}D + \overline{A}BC + ACD + A\overline{B}\overline{D}$.

(C) Right way: f = $\overline{A}D + CD + AB\overline{D}$.

$f = \overline{A}D + CD + A\overline{B}\overline{D}$

Fig. 5-22. Techniques of making loops.

93

of ORed terms. Two of these groups have only two variables (four 1s in each group eliminating two variables each) and one group has three variables. Note that this function is greatly simplified over that of Fig. 5-22B.

5-7. INVERSION BY K-MAP

The K-map can be used to find the inversion of a function. In Fig. 5-23A, f = AB. If you were to change all 1s to 0s and all 0s to 1s, you would have a direct plot of inverted f, or $\bar{f}$. However, this is not necessary if you simply loop all the blank squares (assumed to be 0) as shown by Fig. 5-23B.

If f is plotted on the map of Fig. 5-23C, then $\bar{f}$ is found by looping the blank squares as illustrated by the drawing. To plot $\bar{f}$ directly on the K-map, 1s would be plotted on the map as shown by Fig. 5-23D.

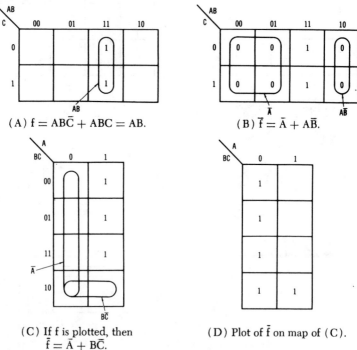

(A) $f = AB\bar{C} + ABC = AB$.

(B) $\bar{f} = \bar{A} + A\bar{B}$.

(C) If f is plotted, then $\bar{f} = \bar{A} + B\bar{C}$.

(D) Plot of $\bar{f}$ on map of (C).

Fig. 5-23. Inversion of function by Karnaugh map.

5-8. THE K-MAP AND REDUNDANCIES

Very often in digital circuit design, the total possible combinations of variables are more than are required in the operation. An example of this is the "decade counter" which counts from 0 to 9 (ten digits) then resets to 0 on the tenth count. This is termed a *binary coded decimal* (bcd) counter, which requires four bits for the necessary count. Review Table 1-2 (Chapter 1) and note that four bits have a maximum capacity of decimal 15; therefore all bits following decimal 9 are redundant in this application.

Redundant combinations may be used on a K-map in the simplification procedure. This is done by marking the squares corresponding to the redundant bits with an "X." Any of these marked squares may then be combined with other squares on the map marked with a 1 whenever so doing will result in further simplification.

For example, assume a decade counter using the four variables (A, B, C, and D) is involved in a problem. This problem says that a binary 1 is to be provided each time the counter reaches the 5 or 7 state for operation of certain associated equipment. Thus the function is:

$$\left. \begin{array}{l} 5 = 0101 = \bar{A}B\bar{C}D \\ 7 = 0111 = \bar{A}BCD \end{array} \right\} = \bar{A}BD$$

(This is *not* the final simplification.)

The problem is plotted as in Fig. 5-24A. The 1s are placed in the squares corresponding to 0101 and 0111. The first simplification becomes $f = \bar{A}BD$.

Now study Fig. 5-24B. The Xs are placed in all squares corresponding to the redundancies, which are 1010 (decimal 10), 1011 (decimal 11), 1100 (decimal 12), 1101 (decimal 13), 1110 (decimal 14) and 1111 (decimal 15).

In this example, two redundancy squares can be included with those squares having a 1 to further simplify the function. The result is four squares within the loop, and the final function is simply BD. Only use the redundant (X) squares if they aid in the simplification.

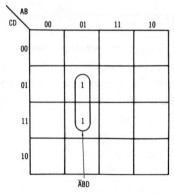

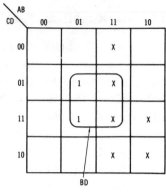

(A) A 1 is provided each time the counter reaches the 5 or 7 state.

(B) Here X denotes redundancy.

Fig. 5-24. Using redundancies to simplify counter function.

EXERCISES

5-1. Give the original terms for which the K-map of Fig. 5-21 shows the solution.

5-2. Solve the Karnaugh map of Fig. 5-25.

5-3. Invert $A\bar{B} + \bar{A}B$ by using a K-map.

5-4. The truth table for two separate functions, f_1 and f_2, is given in Table 5-3. Write the minterm expression for each function in Boolean form.

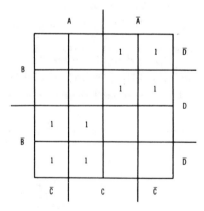

Fig. 5-25. K-map for Exercise 5-2.

5-5. Simplify the function f_1 of Table 5-3 by means of a Karnaugh map.

5-6. Simplify the function f_2 of Table 5-3 by means of a Karnaugh map.

Table 5-3. Truth Table for Functions f_1 and f_2

Decimal	A	B	C	D	f_1	f_2
0	0	0	0	0	0	1
1	0	0	0	1	0	0
2	0	0	1	0	1	1
3	0	0	1	1	1	0
4	0	1	0	0	1	1
5	0	1	0	1	1	1
6	0	1	1	0	1	0
7	0	1	1	1	1	0
8	1	0	0	0	0	1
9	1	0	0	1	0	1
10	1	0	1	0	0	0
11	1	0	1	1	0	0
12	1	1	0	0	1	0
13	1	1	0	1	0	0
14	1	1	1	0	1	0
15	1	1	1	1	0	0

5-7. Plot $f = AB + \overline{A}\overline{B}C + A\overline{B}\overline{D}$ on a K-map.

5-8. What is the simplest Boolean relationship that will produce a 1 only when a decade counter is in the 9 position? Give: (a) redundancies, (b) the function to be solved, and (c) a K-map to solve the problem.

5-9. Simplify $f = AB\overline{C}D + ABCD + ABC\overline{D}$ with a K-map.

5-10. Why should three adjacent 1s be broken down into two 2-loops rather than into a single 2-loop and a single 1?

Practice Problems in Boolean Algebra

This chapter will summarize the areas covered by previous chapters, and provide you will a good basic background in using the Boolean technique. First, 30 problems will be stated, and then their solutions will be given.

6-1. PROBLEMS

Problem 1. Simplify $f=ABC+AB\bar{C}+A\bar{B}C+\bar{A}BC+\bar{A}B\bar{C}$ by (a) the chart method of Section 3-2 and (b) double-check by the K-map method.

Problem 2. Give the function (f) of the following statements by means of truth tables. Let:

A = "Circuit A is working."
B = "Circuit B is working."
S_1 = Statement 1.
S_2 = Statement 2.
S_3 = Statement 3.

Statement 1. "It is not true that circuit A and circuit B are both working."

Statement 2. "Circuit A is not working and circuit B is not working."

Statement 3. "It is not true that circuit A or circuit B is working."

Problem 3. Simplify $f = (\overline{A+B})(BC)$ by (a) algebraic means, and (b) double-check with truth table.

Problem 4. (a) Design a circuit that will produce an output every time a four-bit binary decimal counter is in the 6, 7, 8, or 9 positions. (b) Sketch the logical diagram of the simplified circuit.

Problem 5. Can the function $f = ABC + A\overline{B}\overline{C} + \overline{A}\overline{B}C + \overline{A}BC$ be simplified?

Problem 6. Simplify $f = A + (BC+A)C$ using Boolean algebra. Show each step in the solution.

Problem 7. Simplify $f = (A+B)\overline{AB}$ using Boolean algebra. Show each step in the solution.

Problem 8. Simplify $f = AB(A+B)$ by (a) truth table and (b) Venn diagram.

Problem 9. Using Fig. 4-2B (Chapter 4), describe the class $\overline{AB}$ (NOT A AND B).

Problem 10. Using complements, solve the following: (a) 1111 −1100 and (b) 101000−001100.

Problem 11. Convert decimal 30 to binary notation.

Problem 12. Convert decimal 368.75 to binary notation.

Problem 13. Invert f=AB by using a Karnaugh map.

Problem 14. Reduce the following statement to Boolean form: "It is Tuesday and 2 + 2 is equal to 5."

Problem 15. Reduce the following sentence to Boolean form, and specify the phrase of the sentence for which each symbol stands: "A mouse trap is not suitable for an elephant, or some scientists are going to be very surprised and some men will wish they were mice."

Problem 16. Is the answer as derived directly from Problem 15 in its simplest form? See if you can find the simplest form. (The answer in this book is given by a Karnaugh map.)

Problem 17. Suppose that you have a circuit that provides an INHIBIT function ($f=A\overline{B}$) ORed with the B input, followed by a NOR function. (Remember that NOR implies inversion.) The out-

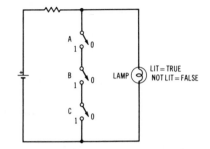

Fig. 6-1. Circuit for Problem 19.

LIT = TRUE
NOT LIT = FALSE

put expression then is $f = \overline{(A\overline{B})} + B$. Simplify this expression. (Hint: This is most readily done by the algebraic method.)

Problem 18. If necessary, review the full binary addition techniques in Chapter 1 (Table 1-8). Draw two Karnaugh maps as follows: (a) for binary sum (S = 1) and (b) for carry output ($C_n = 1$). One of the two can be further simplified. Which one? Do this simplification on the map.

Problem 19. Define the type of operation illustrated by the circuit of Fig. 6-1. Draw the symbolic circuit for this function.

Problem 20. Specify the function (f) for Fig. 6-2.

Problem 21. Simplify the expression obtained in Problem 20 and draw the simplified circuitry. (In this book, the expression is simplified by a truth table.)

Problem 22. "A dog has four legs or a horse has five legs." Is this statement true or false?

Problem 23. "A dog has four legs and a horse has five legs." Is this statement true or false?

Problem 24. "One plus one equals two or the sky is blue." Is this statement true or false? (Be careful of this one!)

Problem 25. Fig. 6-3 shows a logic circuit for $f = AB + AC$. How can you simplify this circuit? (Hint: Distributive rule.)

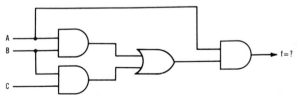

Fig. 6-2. Circuit for Problem 20.

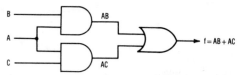

Fig. 6-3. Circuit for Problem 25.

Problem 26. Fig. 6-4A shows the logic circuit for $f=(A+B)$ $(A+C)$. Fig. 6-4B illustrates the conventional algebraic absorptive rule which involves expansion by cross multiplication. Use the absorptive rule to design a logic circuit simpler than that of Fig. 6-4A but which gives the same result.

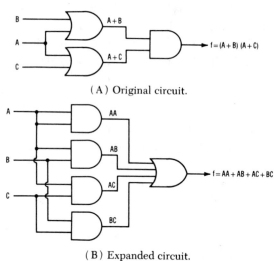

(A) Original circuit.

(B) Expanded circuit.

Fig. 6-4. Circuits for Problem 26.

Problem 27. Draw a logic circuit that will apply high voltage to the final stage of a transmitter if:

A. The filaments are on and time delay contacts are closed.
B. Blowers are on.
C. Door interlocks are closed.

But not if:

D. Exciter is out of frequency limits.
E. Insufficient excitation is present to the final stage.

F. Final stage plate currents are excessive when power applied.

G. Vswr is excessive when power applied.

H. High-voltage power supply is out of limits.

Problem 28. (a) Draw the logic diagram for $f = ACD + A\bar{B}C$ $+ AC\bar{E}$. (b) Factor the expression above by conventional algebraic means and draw the resulting logic diagram. Assume that all variables are available in both true and inverted forms.

Problem 29. (a) Draw the logic diagram for $f = A\bar{B}C + AB\bar{C} + D$. (b) Factor the expression above and draw the resulting logic diagram. (c) What conclusion can you reach?

Problem 30. Can $f = \bar{A}BCD + ABC\bar{D} + A\bar{B}\bar{C}D + \bar{A}B\bar{C}\bar{D} +$ ABCD be simplified? If so, what is the simplified function?

6-2. SOLUTIONS OF PROBLEMS

Problem 1.

(a) See Chart 6-1 for chart solution. Review the technique described in Section 3-2.

Chart 6-1. Chart Form of $f = B + AC$

1	2	3	4	5	6	7	
A	B	C	AB	AC	BC	ABC	
$\bar{A}$	$\bar{B}$	$\bar{C}$	$\bar{A}\bar{B}$	$\bar{A}\bar{C}$	$\bar{B}\bar{C}$	$\bar{A}\bar{B}\bar{C}$	Row 1
$\bar{A}$	$\bar{B}$	C	$\bar{A}\bar{B}$	$\bar{A}C$	$\bar{B}C$	$\bar{A}\bar{B}C$	Row 2
$\bar{A}$	$\boxed{B}$	$\bar{C}$	$\bar{A}B$	$\bar{A}\bar{C}$	$B\bar{C}$	$\bar{A}B\bar{C}$	Row 3
$\bar{A}$	$\boxed{B}$	C	$\bar{A}B$	$\bar{A}C$	BC	$\bar{A}BC$	Row 4
A	$\bar{B}$	$\bar{C}$	$A\bar{B}$	$A\bar{C}$	$\bar{B}\bar{C}$	$A\bar{B}\bar{C}$	Row 5
A	$\bar{B}$	C	$A\bar{B}$	$\boxed{AC}$	$\bar{B}C$	$A\bar{B}C$	Row 6
A	$\boxed{B}$	$\bar{C}$	AB	$A\bar{C}$	$B\bar{C}$	$AB\bar{C}$	Row 7
A	$\boxed{B}$	C	AB	$\boxed{AC}$	BC	ABC	Row 8

(b) Construct the three-variable K-map of Fig. 6-5 and insert the minterm values in the lower right-hand corner of each square. (Review Fig. 5-6 and Table 5-2.) Then it is easy to insert the 1s to match the terms of the problem. For example, ABC =

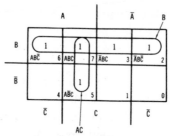

Fig. 6-5. Veitch form of Karnaugh map.

$4+2+1 = 7$, $AB\overline{C} = 4+2+0 = 6$, etc. The horizontal coupling (group) across the entire top row is equal to B, since $A+\overline{A} = 1$ and $C+\overline{C} = 1$. Thus the simplified term is independent of A and C. The vertical coupling of squares 5 and 7 gives AC since $B+\overline{B} = 1$ and therefore the term is independent of B.

Problem 2.

For S_1, see Fig. 6-6A ($S_1 = \overline{AB}$).
For S_2, see Fig. 6-6B ($S_2 = \overline{A}\overline{B}$).
For S_3, see Fig. 6-6C ($S_3 = \overline{A+B} = \overline{A}\overline{B} = S_2$).

Problem 3.

(a) $(\overline{A+B})BC = (\overline{A}\overline{B})BC = \overline{A}(\overline{B}B)C$
$\qquad\qquad\qquad = \overline{A}\cdot 0 \cdot C \qquad\qquad (\overline{B}B = 0)$
$\qquad\qquad\qquad = 0$

(b) See Table 6-1. Note that the terms $\overline{A+B}$ and BC do not reach highs simultaneously; therefore f=0.

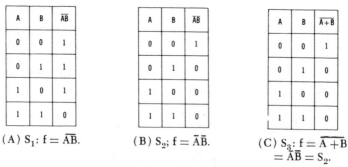

A	B	$\overline{AB}$
0	0	1
0	1	1
1	0	1
1	1	0

(A) S_1: $f = \overline{AB}$.

A	B	$\overline{A}\overline{B}$
0	0	1
0	1	0
1	0	0
1	1	0

(B) S_2; $f = \overline{A}\overline{B}$.

A	B	$\overline{A+B}$
0	0	1
0	1	0
1	0	0
1	1	0

(C) S_3: $f = \overline{A+B}$
$= \overline{A}\overline{B} = S_2$.

Fig. 6-6. Solution of Problem 2.

Table 6-1. Solution of Problem 3(b)

A	B	C	A + B	$\overline{A+B}$	BC	$(\overline{A + B})BC$
0	0	0	0	1	0	0
0	0	1	0	1	0	0
0	1	0	1	0	0	0
0	1	1	1	0	1	0
1	0	0	1	0	0	0
1	0	1	1	0	0	0
1	1	0	1	0	0	0
1	1	1	1	0	1	0

Problem 4.

Redundancies: 1010, 1011, 1100, 1101, 1110, 1111.

Function to be solved: Produce a "1" for 0110, 0111, 1000, 1001.

(a) See Karnaugh map, Fig. 6-7A. Note that one 8-loop is possible to result in a single variable.

(b) See Fig. 6-7B.

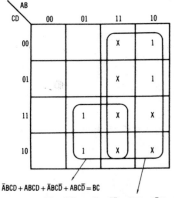

$\overline{A}BCD + ABCD + \overline{A}BC\overline{D} + ABC\overline{D} = BC$

$ABC\overline{D} + A\overline{B}C\overline{D} + AB\overline{C}\overline{D} + A\overline{B}\overline{C}\overline{D} + ABCD + A\overline{B}CD + AB\overline{C}D + A\overline{B}\overline{C}D = A$

$f = A + BC$ (FINAL SIMPLIFICATION)

(A) Karnaugh map.

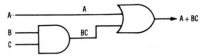

(B) Schematic of simplified circuit.

Fig. 6-7. Solution of Problem 4.

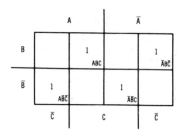

Fig. 6-8. Solution of Problem 5.

Problem 5.

The easiest way to tell if a function can be simplified is to plot it on a Karnaugh map. See Fig. 6-8. Since there are no adjacent 1s, this function cannot be further simplified.

Problem 6.

$$
\begin{aligned}
A+(BC+A)C &= A+BCC+AC \\
&= A+BC+AC \\
&= A+AC+BC \\
&= A(1+C)+BC \\
&= A+BC
\end{aligned}
$$

Problem 7.

$$
\begin{aligned}
(A+B)\overline{AB} &= (A+B)(\overline{A}+\overline{B}) \\
&= A(\overline{A}+\overline{B})+B(\overline{A}+\overline{B}) \\
&= A\overline{A}+A\overline{B}+B\overline{A}+B\overline{B} \\
&= A\overline{B}+B\overline{A} \\
&= A \oplus B \ (\text{exclusive OR})
\end{aligned}
$$

A	B	AB	A + B	AB(A + B)
0	0	0	0	0
0	1	0	1	0
1	0	0	1	0
1	1	1	1	1

IDENTICAL

(A) Truth table showing
AB(A + B) = AB.

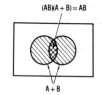

(B) Venn diagram crosshatched
area is AB.

Fig. 6-9. Solution of Problem 8.

Problem 8.

See Fig. 6-9. Remember the following basic rules for Venn diagrams: (1) If the expressions are connected by a "·," the crosshatched area is the area of interest. (2) If the expressions are connected by a "+," the total shaded area and crosshatched area is the area of interest.

Problem 9.

$$\overline{AB} = A\overline{B} \text{ (area 1)} + \overline{A}B \text{ (area 2)} + \overline{A}\overline{B} \text{ (area 3)}$$
$$= A\overline{B} + \overline{A}B + \overline{A}\overline{B}$$
$$= \overline{A}(B+\overline{B}) + \overline{B}(A+\overline{A})$$
$$= \overline{A}+\overline{B} \qquad\qquad (B\overline{B} = 0 \text{ and } A\overline{A} = 0)$$

This simply proves that any class can be represented by either of two expressions (DeMorgan's theorem).

Problem 10.

(a) $1111 - 1100$ $(15-12)$

```
       0011   (inverted: ones complement)
        +1
       0100   (twos complement)
  add 1111
     10011 = +0011 = +3  (answer)
```

(b) $101000-001100$ $(40-12)$

```
      110011   (inverted: ones complement)
         +1
      110100   (twos complement)
 add 101000
    1011100 = +11100 = +28   (answer)
```

Problem 11.

Decimal No.	÷ Base 2	= Result	Remainder (Binary)	
30	÷ 2	= 15	0	
15	÷ 2	= 7	1	
7	÷ 2	= 3	1	Read up
3	÷ 2	= 1	1	
1	÷ 2	= 0	1	

256	128	64	32	16	8	4	2	1	0.5	0.25	0.125
1	0	1	1	1	0	0	0	0	1	1	0

1.	368.75	Number to be converted.
2.	−256.00	Subtract largest whole power of 2. Place a 1 under 256.
3.	112.75	Remainder
4.	−64.00	Subtract largest whole power of 2. Place a 1 under 64.
5.	48.75	Remainder
6.	−32.00	Subtract largest whole power of 2. Place a 1 under 32.
7.	16.75	Remainder
8.	−16.00	Subtract largest whole power of 2. Place a 1 under 16.
9.	0.75	Remainder
10.	−0.50	Subtract largest whole power of 2. Place a 1 under 0.5.
11.	0.25	Remainder
12.	−0.25	Subtract largest whole power of 2. Place a 1 under 0.25.
13.	0.00	Remainder

Double check: Binary 1 0 1 1 1 0 0 0 . 1 1

$$256 + 64 + 32 + 16 + \quad 0.5 + 0.25 = 368.75 \text{ decimal}$$

Fig. 6-10. Solution of Problem 12.

Double-check: 11110 binary $= 16+8+4+2+0 = 30$

Problem 12.

See Fig. 6-10.

Problem 13.

See Fig. 6-11. Note that AB inverted $(\overline{AB})$ is equal to $\bar{A}+\bar{B}$. This is DeMorgan's theorem (Rule 25, Table 3-2).

Problem 14.

$f = A \cdot 0 = 0$. Let A = "It is Tuesday" and B = "2+2=5." The AND statement is false whether or not it is Tuesday, since "2+2= 5" is false.

Problem 15.

A = "A mouse trap is suitable for an elephant."
B = "Some scientists are going to be very surprised."

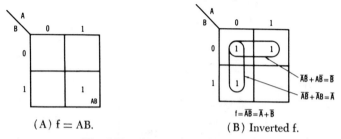

(A) $f = AB$.　　(B) Inverted f.

Fig. 6-11. Solution of Problem 13.

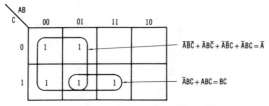

Fig. 6-12. Solution of Problem 16.

C = "Some men will wish they were mice."
Answer: $f = \bar{A} + BC$.

Problem 16.

The function f is in its simplest form as $f = \bar{A} + BC$. See Fig. 6-12.

Problem 17.

$$
\begin{aligned}
\overline{(A\bar{B})+B} &= (\overline{A\bar{B}})\bar{B} \\
&= (\bar{A}+B)\bar{B} \\
&= \bar{A}\bar{B}+B\bar{B} \\
&= \bar{A}\bar{B}+0 \\
&= \bar{A}\bar{B} \\
&= \overline{A+B}
\end{aligned}
$$

Either $\bar{A}\bar{B}$ or $\overline{A+B}$ is the correct answer.

Problem 18.

See Fig. 6-13. Note that the sum expression cannot be further simplified (although other sum expressions can exist), but the carry out expression (C_n) can be simplified. Thus $C_n = \bar{A}BC+A\bar{B}C+AB\bar{C}+ABC = AB+AC+BC$.

Problem 19.

Switches A, B, and C are true when closed, and false when open. The lamp is true when lit and false when not lit. The lamp will be lit (f=true) when any one or more than one of the switches is open. The lamp will not be lit (f=false) when and only when all switches are closed. This is a NAND (NOT AND) function. The symbol for this particular circuit is shown in Fig. 6-14.

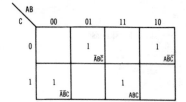

(A) Sum $\overline{A}\overline{B}C + \overline{A}B\overline{C} + A\overline{B}\overline{C} + ABC$ cannot be further simplified.

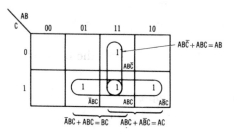

$C_n = AB + AC + BC$
(FINAL SIMPLIFICATION)

(B) Carry out $C_n = \overline{A}BC + A\overline{B}C + AB\overline{C} + ABC$.

Fig. 6-13. Solution of Problem 18.

Problem 20.

$f = A(AB+BC)$. See Fig. 6-15.

Problem 21.

See Table 6-2. $A(AB+BC) = AB$ (answer).

Table 6-2. Truth Table Shows That A(AB + BC) = AB

A	B	C	AB	BC	AB + BC	A(AB + BC)
0	0	0	0	0	0	0
0	0	1	0	0	0	0
0	1	0	0	0	0	0
0	1	1	0	1	1	0
1	0	0	0	0	0	0
1	0	1	0	0	0	0
1	1	0	1	0	1	1
1	1	1	1	1	1	1

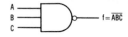

Fig. 6-14. Solution of Problem 19.

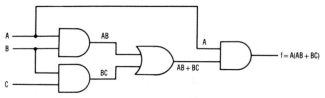

Fig. 6-15. Solution of Problem 20.

Problem 22.

This is a true statement because of the operator OR, regardless of whether the OR is inclusive or exclusive. With the connective OR, if only one of the propositions is true and the other false, the combined statement is true.

Problem 23.

This is the AND operation, so both parts of the statement would need to be true for the combined sentence to be true. Thus this statement is false.

Problem 24.

Depends upon whether you are considering the inclusive or exclusive OR. It is a true statement if either or both propositions are true if you have an inclusive OR (one or both). It is false if both propositions are true and you have an exclusive OR (either one but not both).

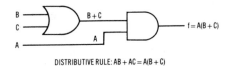

DISTRIBUTIVE RULE: AB + AC = A(B + C)

Fig. 6-16. Solution of Problem 25.

Problem 25.

See Fig. 6-16. This is the Distributive Rule (Rule 21, Table 3-2). It is similar to algebraic factoring.

111

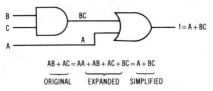

$$AB + AC = AA + AB + AC + BC = A + BC$$

ORIGINAL   EXPANDED   SIMPLIFIED

Fig. 6-17. Solution of Problem 26.

Problem 26.

See Fig. 6-17. Since the expanded equation has AA as one of the OR terms, and AA=A, then the output is high (logical 1) whenever A is present, and the terms AB and AC are redundant.

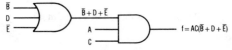

Fig. 6-18. Solution of Problem 27.

Problem 27.

See Fig. 6-18. The Boolean expression is $f=ABC\overline{(D+E+F+G+\overline{H})}$. It is assumed that a logical 1 exists for all variables in the

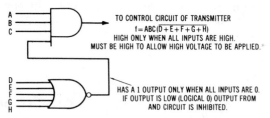

(A) Circuit without factoring.

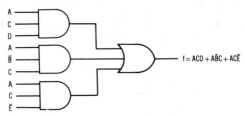

FACTORING: $ACD + A\overline{B}C + AC\overline{E} = AC(\overline{B} + D + \overline{E})$

(B) Simplified (factored) circuit requires half the circuitry.

Fig. 6-19. Solution of Problem 28.

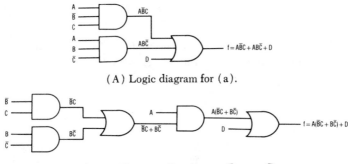

(A) Logic diagram for (a).

(B) Factoring: $A\bar{B}C + AB\bar{C} + D = A(\bar{B}C + B\bar{C}) + D$.

Fig. 6-20. Solution of Problem 29.

table when the state condition exists. Thus if any of the variables D through H is high (indicating a fault), the output of the NOR circuit will be 0 and inhibit the output from the AND circuit, preventing application of high voltage.

Problem 28.

See Fig. 6-19.

Problem 29.

See Fig. 6-20. (c) In this example, the factored form is more costly than the unfactored form (just the opposite of Problem 28). You can conclude that both factored and unfactored forms should be tried before deciding on the logic circuitry for any particular application.

Fig. 6-21. Solution of Problem 30.

AB CD	00	01	11	10
00		1 $\bar{A}B\bar{C}\bar{D}$		
01				1 $A\bar{B}\bar{C}D$
11		1 $\bar{A}BCD$		
10	1 $\bar{A}\bar{B}C\bar{D}$		1 $ABC\bar{D}$	

113

Problem 30.

See Fig. 6-21. The expression cannot be further simplified since there are no adjacent 1s on the Karnaugh map.

Answers to Exercises

CHAPTER 1

1-1. 0 volts = logical 1, and −4 volts = logical 0.

1-2. AND, OR, and NOT.

1-3. NOT means the inverse, or opposite, or the stated condition. Thus if B = 1, B̄ = 0 (read if B=1, B NOT = 0).

1-4. Read A AND B OR A NOT AND B NOT.

1-5. Exercise 1-4 says that a logical 1 output occurs only when A AND B OR A NOT AND B NOT occur. If *any other* condition exists, such as A NOT AND B OR A AND B NOT, the output is logical 0.

1-6. (a) 32+8+1+0.25 = 41.25
(b) 8+1+0.5+0.25+0.125+0.0625 = 11.9375
(c) 128+64+32+8+2+0.5+0.125 = 234.625
(d) 1+0.0625 = 1.0625

1-7. (a)   1000   (8)
   + 1001   (9)
     10001   (17)
(b)   1101.100   (13.5)
  + 1000.101 (8.625)
    10110.001 (22.125)
(c)   0000.1110 (0.875)
 + 0001.1110 (1.875)
   0010.1100 (2.75)
(d)   0000.11111 (0.96875) Note: 2^{-5} = 0.03125
 + 0011.11100 (3.875)
   0100.11011 (4.84375)

1-8. (a) arithmetical:

$$
\begin{array}{r}
1110 \\
-0110 \\
\hline
1000
\end{array}
$$

complement:

$$
\begin{array}{r}
1001 \\
+\quad 1 \\
\hline
1010 \\
\text{add } 1110 \\
\hline
11000
\end{array}
$$

(inverted 0110: ones complement)

(complement of 0110: twos complement)

(answer) $= +8$

(b) arithmetical:

$$
\begin{array}{r}
101001 \\
-001010 \\
\hline
011111
\end{array}
$$

(answer)

Procedure for above is as follows, starting with lsd as 1st digit.

1st digit: $1-0 = 1$.

2nd digit: $0-1 =$ difference of 1 and borrow 1.

3rd digit: no 1 exists to borrow yet, so 0 in top row becomes 1 and $1-0 = 1$ with borrow 1 carried over.

4th digit: now a 1 exists to borrow, so 1 in top row becomes 0 and $0-1 =$ difference of 1 and borrow 1.

5th digit: no 1 to borrow, so 0 top row becomes 1 and $1-0 = 1$ with borrow 1 carry over.

6th digit: now a 1 exists to borrow, so 1 in top row becomes 0 and $0-0 = 0$. Prove answer by noting that decimal equivalent is $41-10 = 31$. Note that this procedure is cumbersome relative to the following *complementary* operation:

(b) complement:

$$
\begin{array}{r}
110101 \\
+\qquad 1 \\
\hline
110110 \\
\text{add } \quad 101001 \\
\hline
1011111
\end{array}
$$

(inverted 001010: ones complement)

(complement of 001010: twos complement)

$= +011111 = 31$

CHAPTER 2

2-1. (a) "A AND B," (b) "NOT A AND B," (c) "A NOT AND B NOT," (d) "A OR B inclusive" (either or both), (e) NOT A OR B (inverted A+B), (f) "A OR B exclusive" (either/or but not both).

2-2. $A \oplus B = \bar{A}B + A\bar{B}$

2-3. "A NOT AND B OR A AND B NOT."

2-4. See Table A-1 for a four-input AND gate. Note that the truth table starts with binary 0000 and increases by adding 1 each time until the number 15 (1111) is reached. This ensures that no combination is omitted.

Table A-1. Truth Tables for f = ABCD

A	B	C	D	f
0	0	0	0	0
0	0	0	1	0
0	0	1	0	0
0	0	1	1	0
0	1	0	0	0
0	1	0	1	0
0	1	1	0	0
0	1	1	1	0
1	0	0	0	0
1	0	0	1	0
1	0	1	0	0
1	0	1	1	0
1	1	0	0	0
1	1	0	1	0
1	1	1	0	0
1	1	1	1	1

2-5. $A+\bar{B}+\bar{C}+D$.

2-6. See Table A-2. For clarity, the value of $\bar{B}$ is included where if $B=1$, $\bar{B}=0$, and if $B=0$, $\bar{B}=1$. This shows that an output (logical 1) occurs for all conditions except where $A=0$ AND $B=1$.

Table A-2. Solution of Exercise 2-6

A	B	$\bar{B}$	$A + \bar{B}$
0	0	1	1
0	1	0	0
1	0	1	1
1	1	0	1

2-7. See Table A-3. Note that the value of C is placed in the truth table to make the function clear. (If C=1, $\bar{C}$=0, and if C=0, $\bar{C}$=1.) An output (logical 1) occurs only when inputs are AB$\bar{C}$ (A=1, B=1, C=0). In this case $\bar{C}$ = 1, and the AND function (all inputs coincidently high) occurs.

Table A-3. Solution of Exercise 2-7

A	B	C	$\bar{C}$	AB$\bar{C}$
0	0	0	1	0
0	0	1	0	0
0	1	0	1	0
0	1	1	0	0
1	0	0	1	0
1	0	1	0	0
1	1	0	1	1
1	1	1	0	0

2-8. See Table A-4. Note that no output (logical 0) now occurs *only* when A=0, B=0, and C=1 (so that $\bar{C}$ = 0).

Table A-4. Solution of Exercise 2-8

A	B	C	$\bar{C}$	A + B + $\bar{C}$
0	0	0	1	1
0	0	1	0	0
0	1	0	1	1
0	1	1	0	1
1	0	0	1	1
1	0	1	0	1
1	1	0	1	1
1	1	1	0	1

CHAPTER 3

3-1. Let A = Jack is present, B = Jill is present, and C = It is snowing. Then we must have a functional output (f=1) only when Jack and Jill are both present, and it is not snowing, so:

$$f = AB\bar{C}$$

This says that a logical 1 output occurs when A is true, B is true, and C is false (the negation of "it is snowing").

3-2. See Table A-5.

Table A-5. Truth Table for $AB\bar{C}$

A	B	C	$\bar{C}$	$AB\bar{C}$
0	0	0	1	0
0	0	1	0	0
0	1	0	1	0
0	1	1	0	0
1	0	0	1	0
1	0	1	0	0
1	1	0	1	1
1	1	1	0	0

3-3. The expression $AB0$ can be written as $A(\bar{B}0)$. Since the quantity $(\bar{B}0) = 0$, then A AND 0 $= 0$. Anytime a zero is in the AND operation, the result is 0.

3-4. $A+A+\bar{A}+1 = A+\bar{A}+1$ (since $A+A=A$)
                 $= 1+1$ (since $A+\bar{A}=1$)
                 $= 1$

3-5. See Fig. A-1. Since $\bar{A}+A=1$, the functional output (f_3) is B. Note that just a straight wire from B is all that is necessary.

3-6. See Fig. A-2. The output is $f_6 = \bar{A}$.

3-7. See Fig. A-3. Just one simple inverter stage is all that is necessary.

3-8. Rule 23 states $A+\bar{A}B = A+B$. Reasoning aloud is equivalent to using a truth table. If $A=1$, then $\bar{A}=0$, and $A+\bar{A}B = 1+0B = 1$. Also, $A+B$ is $1+B = 1$. So now we know that if $A=1$, the

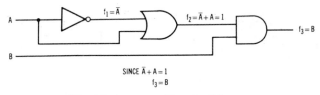

Fig. A-1. Answer to Exercise 3-5.

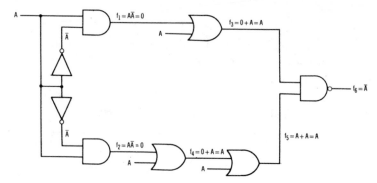

Fig. A-2. Answer to Exercise 3-6.

equality holds. Now assume $A=0$; then $\bar{A}=1$ and $A+\bar{A}B = 0+1B = B$. Then $A+B = 0+B = B$. Again the equality holds.

3-9. $\overline{A+B+C} = (\overline{A+B})\bar{C} = (\bar{\bar{A}}\bar{B})\bar{C} = \bar{\bar{A}}\bar{B}\bar{C}$, but $\bar{\bar{A}}=A$, so $\bar{\bar{A}}\bar{B}\bar{C} = A\bar{B}\bar{C}$.

3-10. $f=AC+AD+BC+BD$
$\quad =A(C+D) + B(C+D)$ Rule 21
$\quad =(C+D)(A+B)$ Rule 21

Fig. A-3. Equivalent circuit for Figs. 3-10 and A-2.

CHAPTER 4

4-1. See Fig. A-4.

4-2. See Fig. A-5.

4-3. See Fig. A-6.

4-4. See Fig. A-7. Class B need not be included since the expression contains no B term.

$f=A\bar{B}$ OR $B\bar{A}$
Fig. A-4. Answer to Exercise 4-1.

$f = AB + \bar{A}\bar{B}$
Fig. A-5. Answers to Exercise 4-2.

$f = AB + C$

Fig. A-6. Answer to Exercise 4-3.

$f = A + \bar{C}$

Fig. A-7. Answers to Exercise 4-4.

4-5. See Fig. A-8. Class B need not be included.

4-6. See Fig. A-9.

$f = AB\bar{C}$

Fig. A-8. Answer to Exercise 4-5.

$f = AC$

Fig. A-9. Answer to Exercise 4-6.

4-7. (a) $\overline{AB} + B\bar{A} = (\bar{A} + \bar{B}) + B\bar{A}$
$$= \bar{A}(1 + B) + \bar{B}$$
$$= \bar{A}(1) + \bar{B}$$
$$= \bar{A} + \bar{B}$$
$$= \overline{AB}$$

(b) See drawing of OR subclass, Fig. 4-6, row 4.

4-8. (a) $A\bar{B} + B\bar{A} + \overline{AB} = A\bar{B} + B\bar{A} + \bar{A} + \bar{B}$
$$= \bar{A} + \bar{A}B + \bar{B} + \bar{B}A$$
$$= \bar{A}(1 + B) + \bar{B}(1 + A)$$
$$= \bar{A} + \bar{B}$$
$$= \overline{AB}$$

(b) See Fig. 4-6, row 4, OR subclass.

4-9. See Table A-6.

4-10. See Table A-7.

CHAPTER 5

5-1. $\bar{A}\bar{B}\bar{C}\bar{D} + A\bar{B}\bar{C}\bar{D} + \bar{A}B\bar{C}D + AB\bar{C}D + \bar{A}BCD + ABCD +$
$\bar{A}\bar{B}C\bar{D} + A\bar{B}C\bar{D}$.

5-2. See Fig. A-10.

Table A-6. Proof That $\bar{A}B\bar{C} + \bar{A}BC + ABC + AB\bar{C} = B$

A	B	C	Ā	B̄	C̄	ĀBC̄	ĀBC	ABC	ABC̄	ĀBC̄ + ĀBC + ABC + ABC̄
0	0	0	1	1	1	0	0	0	0	0
0	0	1	1	1	0	0	0	0	0	0
0	1	0	1	0	1	1	0	0	0	1
0	1	1	1	0	0	0	1	0	0	1
1	0	0	0	1	1	0	0	0	0	0
1	0	1	0	1	0	0	0	0	0	0
1	1	0	0	0	1	0	0	0	1	1
1	1	1	0	0	0	0	0	0	0	1

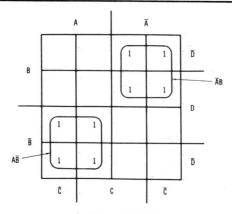

$f = A\bar{B} + \bar{A}B$ (FINAL SIMPLIFICATION)

Fig. A-10. Answer to Exercise 5-2.

Table A-7. Proof That $\bar{A}\bar{B}C + A\bar{B}C + ABC + \bar{A}BC = C$

A	B	C	Ā	B̄	C̄	ĀB̄C	AB̄C	ABC	ĀBC	ĀB̄C + AB̄C + ABC + ĀBC
0	0	0	1	1	1	0	0	0	0	0
0	0	1	1	1	0	1	0	0	0	1
0	1	0	1	0	1	0	0	0	0	0
0	1	1	1	0	0	0	0	0	1	1
1	0	0	0	1	1	0	0	0	0	0
1	0	1	0	1	0	0	1	0	0	1
1	1	0	0	0	1	0	0	0	0	0
1	1	1	0	0	0	0	0	1	0	1

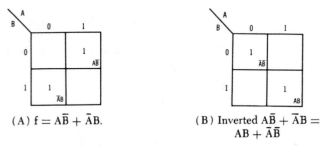

(A) $f = A\bar{B} + \bar{A}B$.

(B) Inverted $A\bar{B} + \bar{A}B = AB + \bar{A}\bar{B}$

Fig. A-11. Answer to Exercise 5-3.

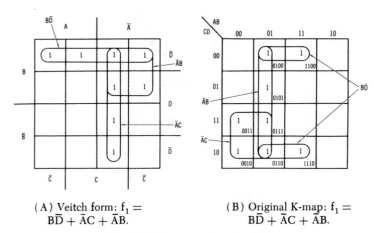

(A) Veitch form: $f_1 = B\bar{D} + \bar{A}C + \bar{A}B$.

(B) Original K-map: $f_1 = B\bar{D} + \bar{A}C + \bar{A}B$.

Fig. A-12. Answer to Exercise 5-5.

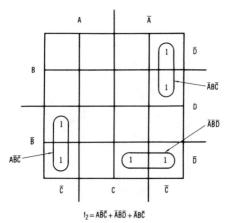

$f_2 = A\bar{B}\bar{C} + \bar{A}\bar{B}\bar{D} + \bar{A}B\bar{C}$

Fig. A-13. Answer to Exercise 5-6.

5-3. See Fig. A-11. Fig. A-11A is the exclusive OR function $A\overline{B}+\overline{A}B$. Fig. A-11B ($A\overline{B}+\overline{A}B$ inverted) is the exclusive NOR function $AB+\overline{A}\overline{B}$. The 1s and 0s are simply interchanged on the K-map.

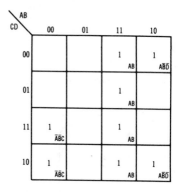

PLOT OF $AB + \overline{A}\overline{B}C + A\overline{B}\overline{D}$

Fig. A-14. Answer to Exercise 5-7.

REDUNDANCIES

$$
\begin{array}{cccccc}
10 & 11 & 12 & 13 & 14 & 15 \\
\downarrow & \downarrow & \downarrow & \downarrow & \downarrow & \downarrow \\
A\overline{B}C\overline{D} & A\overline{B}CD & AB\overline{C}\overline{D} & AB\overline{C}D & ABC\overline{D} & ABCD \\
\downarrow & \downarrow & \downarrow & \downarrow & \downarrow & \downarrow \\
1010 & 1011 & 1100 & 1101 & 1110 & 1111
\end{array}
$$

FUNCTION TO BE SOLVED: PRODUCE A 1 FOR $1001 = A\overline{B}\overline{C}D$

SEE K-MAP BELOW

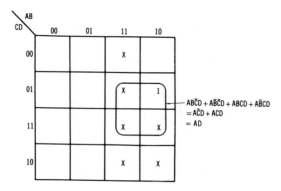

$AB\overline{C}D + A\overline{B}\overline{C}D + ABCD + A\overline{B}CD$
$= A\overline{C}D + ACD$
$= AD$

$f = AD$ (FINAL SIMPLIFICATION)

Fig. A-15. Answer to Exercise 5-8.

5-4. $f_1 = \overset{2}{\overline{A}\overline{B}C\overline{D}} + \overset{3}{\overline{A}\overline{B}CD} + \overset{4}{\overline{A}B\overline{C}\overline{D}} + \overset{5}{AB\overline{C}D} + \overset{6}{\overline{A}BC\overline{D}} + \overset{7}{\overline{A}BCD} +$

$\overset{12}{AB\overline{C}\overline{D}} + \overset{14}{ABC\overline{D}}$

$f_2 = \overset{0}{\overline{A}\overline{B}\overline{C}\overline{D}} + \overset{2}{\overline{A}\overline{B}C\overline{D}} + \overset{4}{\overline{A}B\overline{C}\overline{D}} + \overset{5}{\overline{A}B\overline{C}D} + \overset{8}{A\overline{B}\overline{C}\overline{D}} + \overset{9}{A\overline{B}\overline{C}D}$

5-5. See Fig. A-12, which shows solution on both forms of K-maps.

5-6. See Fig. A-13.

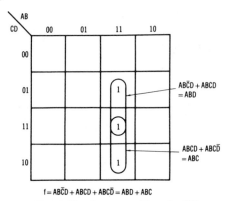

$f = AB\overline{C}D + ABCD + ABC\overline{D} = ABD + ABC$

Fig. A-16. Answer to Exercise 5-9.

5-7. See Fig. A-14.

5-8. See Fig. A-15.

5-9. See Fig. A-16.

5-10. To get the simplest Boolean term. Note that if only a single 2-loop is used, you have one 3-variable term and then one 4-variable term for the single 1. Two 2-loops eliminate one variable in a term.

Index